Environmental Stewardship
in the Judeo-Christian Tradition

Environmental Stewardship in the Judeo-Christian Tradition

Jewish, Catholic, and Protestant Wisdom on the Environment

Environmental Stewardship in the Judeo-Christian Tradition

Cover Image: Our House. © Arnie Mateo. 2006. Image from www.sxc.hu

ISSN 1-880595-15-x
ISSN 978-880595-15-x

Acton Institute for the Study of Religion and Liberty

ACTON INSTITUTE

Ottawa Avenue, NW, Suite 301
Grand Rapids, Michigan 49503
Phone: 616-454-3080
Fax: 616-454-9454
www.acton.org

Printed in the United States of America

Reprinted 2008

Contents

Foreword

The biblical starting point for any discussion of the nature of religious environmental stewardship must begin with the witness of the Book of Genesis: "So God created man in his own image, in the image of God he created him; male and female he created them. And God blessed them, and God said to them, 'Be fruitful and multiply, and fill the earth and subdue it; and have dominion over the fish of the sea and over the birds of the air and over every living thing that moves upon the earth'" (Gen. 1:27–28). In our modern times, however, this biblical vision of the relationship between God, man, and nature is muddled by two false views. The one sees the natural world as the source of all value, man as an intruder, and God, if he exists at all, as so immanent in the natural order that he ceases to be distinguishable from it. The other places man as the source of all values, the natural order as merely instrumental to his aims, and God as often irrelevant.

Genesis presents a radically different picture of how the world is put together. In this account, God is the source of all values—in truth, he is the source of everything, calling it into being out of nothing by his powerful word. Man is part of this order essentially and, what is more, by the virtue of his created nature is placed at the head of creation as its steward. Yet this stewardship can never be arbitrary or anthropocentric, as the old canard goes, for this notion implies that man rules creation in God's stead and must do so according to his divine will.

In light of these contemporary confusions about the true nature of stewardship, and because this concept is so central to the concerns of the Judeo-Christian religious tradition and of the free society, the Acton Institute for the Study of Religion and Liberty has committed herself to articulating a vision of environmental stewardship informed by sound theological reflection, honest scientific inquiry, and rigorous economic thinking. To this end, the Institute brought together twenty-five clergy, theologians, economists, environmental scientists, and policy experts in West Cornwall, Connecticut, October 1999, to discuss the aspects of this problem and to lay the intellectual groundwork for further inquiry.

Out of this important meeting was born the idea of composing an interfaith statement that would express common concerns, beliefs, and aspirations about environmental stewardship. Over the course of months, an early draft was vetted by many of the nation's leading Jewish, Catholic, and Protestant minds, and a final version of the Cornwall Declaration on Environmental Stewardship, which was agreed upon on February 1, 2000.

Since then, the Acton Institute, along with the Interfaith Council for Environmental Stewardship* (a broad-based coalition of individuals and organizations committed to the principles espoused in the Cornwall Declaration), began distributing the declaration and promoting its principles within the religious community. In addition, the Acton Institute, in conjunction with the Interfaith Council, developed a series of accompanying essays contained in this volume. Each essay contains the wisdom of its own tradition, and was created with the help of editorial boards comprised of respected Jewish, Catholic, and Protestant thinkers committed to truth and understanding. These three documents each help flesh out the theoretical content of environmental stewardship and its practical application as outlined in the Cornwall Declaration.

I am proud to present the Cornwall Declaration and these documents in the hope that they will contribute significantly to clarifying and advancing the important contemporary conversation about environmental stewardship, helping us all see our moral and religious responsibilities in keeping and tilling the garden that is our world.

—Father Robert A. Sirico
President, Acton Institute
April 17, 2000

*recently renamed the "Cornwall Alliance for Environmental Stewardship - editor

Introduction

When a diverse group of religious thinkers gathered in West Cornwall, Connecticut in October 1999 to hammer out what is now called the "Cornwall Declaration," the views of religious Americans on the environment was not the fashionable topic it is today. References to "Christian views on the environment" appeared from time to time, but probably the most commonly reported "evangelical" view on environmental issues was the view attributed to former Secretary of the Interior James Watt (from 1981-1983). For years, the story circulated that Watt had claimed that environmental issues didn't really matter, because, "After the last tree is felled, Christ will come back." As it happens, Watt never said any such thing. The closest comment he ever made along these lines was before Congress, but it made the opposite point: "I do not know how many future generations we can count on before the Lord returns, whatever it is we have to manage with a skill to leave the resources needed for future generations."

The Watt myth may have circulated for so long because it fit the stereotype that conservative religious believers cared little for the health of our natural environment. This stereotype goes back at least to Lynn White's famous 1967 article in *Science* magazine, which blamed the biblical view of man's dominion over the environment for present day environmental degradation. Although White's argument has been refuted decisively by religious scholars, his stereotype lives on.

A lot has changed in the seven years since the Cornwall Declaration was published. Two more reports of the UN's Intergovernmental Panel on Climate Change (IPPC) have declared that global warming is probably induced by human activities, Al Gore's documentary *An Inconvenient Truth* has won an Oscar, and Live Earth concerts have broadcast live from every continent on the same weekend. Amidst this flurry of activity, the religious debate over the environment has broken into the mainstream.

There was a glimmer of this with the popular 2002 campaign spearheaded by the Evangelical Environmental Network that asked "What would Jesus Drive?" But the big publicity started in February 2006, when *The New York Times* and other major media outlets reported the Evangelical Climate Initiative. The brief document was signed by a group of 86 evangelical leaders, who announced their support for what *The New York Times* called "a major initiative to fight global warming." As part of the "Evangelical Climate Initiative," they called for "federal legislation that would require reductions in carbon dioxide emissions through 'cost-effective, market-based mechanisms.'" Similar reports have followed the views of Richard Cizik, of the National Association of Evangelicals, who is a strong believer in catastrophic, human-induced climate change. And reports of Pope Benedict's environmental views are now commonplace, if not especially accurate.

Much of the reporting has been shallow and one sided, however, and has served to confuse separate issues. Regrettably, many of the religious leaders who have spoken about the environment have contributed to the confusion.

In fact, most of the positive reporting of Christian "concern" for the environment has focused only on those believers who accept the party line on catastrophic, human induced global warming. They are seamlessly assimilated into the media's relentless campaign to declare certainty and scientific consensus where they don't exist. References to theology barely get beyond useful catch phrases such as "This is God's world and we have to take care of it." And those religious thinkers who question the party line are inaccurately described as anti-environment.

Poor framing of the issue began almost immediately after the Evangelical Climate Initiative (ECI) was launched February 8, 2006. Frank James, in a *Chicago Tribune* story the next day, described the disagreement among evangelical Christians this way:

But environmental issues have proved divisive within the body of believers who identify themselves as evangelicals. Some who believe the world is in the "end times," with a return of Jesus imminent, have not seen the necessity of protecting the environment for the long term. Others, meanwhile, have taken the view espoused by the evangelicals who unveiled their campaign Wednesday, that humans were given dominion over the Earth with the responsibility to protect it.

Notice the false dilemma: If you're an evangelical who agrees with the ECI, then you care about the environment. If you disagree with the ECI, then you don't care about the environment because you're expecting the Lord's return any day now.

One could chalk this up to media bias, except that those who spearheaded the ECI have done little to challenge it themselves. *A fair and honest debate about religious responses to environmental issues should always distinguish theological principles from prudential judgments.*

With respect to the environment, the theological principles are easily stated and uncontroversial. The biblical picture is that human beings, as image bearers of God, are placed as stewards over the created order. We bear a responsibility for how we treat and use it. We are part of the creation, as well as its crowning achievement. God intends for us to use and transform the natural world around us for good purposes. Proper use is not misuse. But as fallen creatures, we can mess things up. No serious thinker in the Judeo-Christian tradition questions these basic principles.

Prudential judgments are another thing entirely. They require careful analysis of the relevant scientific, economic, and political aspects of an issue. They require us to weigh costs and benefits, and to discern where facts leave off and fashion begins.

When it comes to global warming, for instance, there are at least four separate issues to keep in mind. You don't need to be a climate expert to recognize them.

(1) Is the planet warming?

(2) If the planet is warming, is human activity (like CO2 emissions) causing it?

(3) If the planet is warming, and we're causing it, is it bad overall?

(4) If the planet is warming, we're causing it, and it's bad, would the policies commonly advocated (e.g., the Kyoto Protocol, legislative restrictions on CO2 emissions) make any difference, or would their cost exceed their benefit?

To answer (1) and (2), one must consider a wide range of scientific evidence, theorizing, and speculation, drawing on disciplines as diverse as meteorology, astrophysics, geology, and probability theory. And the very nature of the questions and the evidence means answers will always be subject to deep uncertainty.

To answer (3) and (4), one must do careful economic reasoning. As a result, religious people who agree that we are stewards of our environment (the principle) can easily *disagree* on how to answer these questions (the application). It is an uncharitable mistake to treat the answers to these questions as if they hinged on the theological principle.

The problem with the Evangelical Climate Initiative, to take one popular example, is that it treats the answers to these four questions as obviously "yes." And it's only on that assumption that the statement can connect our responsibility as stewards with a specific policy position. Theological principle and prudential judgment have been dangerously confused.

The same confusion was present in the 2002 initiative, sponsored by the Evangelical Environmental Network, which asked: "What would Jesus Drive?" The campaign encouraged participants to advocate very specific and coercive federal regulations. Even ignoring the flippant reference to Jesus, the question seems designed to create moral confusion. The campaign targeted SUVs, since SUVs use more fuel than most smaller cars, and the Chevrolet Corporation—for sponsoring a Christian rock concert. "Through this gospel tour," explained Jim Ball, executive director of the Evangelical Environmental Network, "Chevrolet is promoting certain vehicles that get very low gas mileage and produce significant pollution, harming human health and the rest of God's creation." The subtext, which betrays a stunning lack of moral proportion, was that manufacturing and driving an SUV are unambiguous moral evils.

"What would Jesus drive?" is clever marketing, but terrible moral reasoning. The question doesn't have one right answer. One's choice of transportation, like all prudential judgments, is the outcome of a compromise of conflicting goals, most of which have some value. Fuel economy is only one of dozens of legitimate reasons that people ponder when choosing what, or whether they'll drive. We also consider price, family size, occupation, local geography, quality, weather, safety, lifestyle, other available transportation options, and myriad factors beyond accounting. There's no universal measure for comparing them, so every person will weight them differently. Fuel economy doesn't trump all these other values, especially since some cars (such as hybrids) have better than average fuel economy, but require more energy both to construct and to recycle than other, less fuel efficient cars. So an outside observer can't evaluate the morality of your transportation choice just by observing that you drive an SUV. Such complexities make prudential judgments about transportation qualitatively different from moral evils like selling child pornography or torturing a kitten for the fun of it, which are intrinsically evil.

None of this means religious leaders should avoid public policy and stick to theology. It means that we have a moral obligation to distinguish theological principles from how they are applied in any given instance. The Cornwall Declaration and the accompanying essays in this volume were written to do just that. They were not written to provide theological rationale for current environmentalist fashion. Rather, they seek to articulate the broad Judeo-Christian theological principles concerning the environment, and to distinguish those principles from contrary ideas popular in the environmental movement. For that reason, the Acton Institute, along with the Cornwall Alliance, has decided to reissue these statements, along with the Cornwall Declaration, in the hope of encouraging more serious dialogue about environmental stewardship among all believers in the Judeo-Christian tradition.

—Jay W. Richards, Ph.D.
Research Fellow & Director, Acton Media, Acton Institute
September 10, 2007

Part 1

The Cornwall Declaration on Environmental Stewardship

The past millennium brought unprecedented improvements in human health, nutrition, and life expectancy, especially among those most blessed by political and economic liberty and advances in science and technology. At the dawn of a new millennium, the opportunity exists to build on these advances and to extend them to more of the earth's people.

At the same time, many are concerned that liberty, science, and technology are more a threat to the environment than a blessing to humanity and nature. Out of shared reverence for God and His creation and love for our neighbors, we Jews, Catholics, and Protestants, speaking for ourselves and not officially on behalf of our respective communities, joined by others of good will, and committed to justice and compassion, unite in this declaration of our common concerns, beliefs, and aspirations.

Our Concerns

Human understanding and control of natural processes empower people not only to improve the human condition but also to do great harm to each other, to the earth, and to other creatures. As concerns about the environment have grown in recent decades, the moral necessity of ecological stewardship has become increasingly clear. At the same time, however, certain misconceptions about nature and science, coupled with erroneous theological and anthropological positions, impede the advancement of a sound environmental ethic. In the midst of controversy over such matters, it is critically important to remember that while passion may energize environmental activism, it is reason—including sound theology and sound science—that must guide the decision-making process. We identify three areas of common misunderstanding:

1. *Many people mistakenly view humans as principally consumers and polluters rather than producers and stewards.* Consequently, they ignore our potential, as bearers of God's image, to add to the earth's abundance. The increasing realization of this potential has enabled people in societies blessed with an advanced economy not only to reduce pollution, while producing more of the goods and services responsible for the great improvements in the human condition, but also to alleviate the negative effects of much past pollution. A clean environment is a costly good; consequently, growing affluence, technological innovation, and the application of human and material capital are integral to environmental improvement. The tendency among some to oppose economic progress in the name of environmental stewardship is often sadly self-defeating.

2. *Many people believe that "nature knows best," or that the earth—untouched by human hands—is the ideal.* Such romanticism leads some to deify nature or oppose human dominion over creation. Our position, informed by revelation and confirmed by reason and experience, views human stewardship that unlocks the potential in creation for all the earth's inhabitants as good. Humanity alone of all the created order is capable of developing other resources and can thus enrich creation, so it can properly be said that the human person is the most valuable resource on earth. Human life, therefore, must be cherished and allowed to flourish. The alternative-denying the possibility of beneficial human management of the earth-removes all rationale for environmental stewardship.

3. *While some environmental concerns are well founded and serious, others are without foundation or greatly exaggerated.* Some well-founded concerns focus on human health problems in the developing world arising from inadequate sanitation, widespread use of primitive biomass fuels like wood and dung, and primitive agricultural, industrial, and commercial practices; distorted resource consumption patterns driven by perverse economic incentives; and improper disposal of nuclear and other hazardous

wastes in nations lacking adequate regulatory and legal safeguards. Some unfounded or undue concerns include fears of destructive manmade global warming, overpopulation, and rampant species loss. The real and merely alleged problems differ in the following ways:

   a. The former are proven and well understood, while the latter tend to be speculative.
   b. The former are often localized, while the latter are said to be global and cataclysmic in scope.
   c. The former are of concern to people in developing nations especially, while the latter are of concern mainly to environmentalists in wealthy nations.
   d. The former are of high and firmly established risk to human life and health, while the latter are of very low and largely hypothetical risk.
   e. Solutions proposed to the former are cost effective and maintain proven benefit, while solutions to the latter are unjustifiably costly and of dubious benefit.

Public policies to combat exaggerated risks can dangerously delay or reverse the economic development necessary to improve not only human life but also human stewardship of the environment. The poor, who are most often citizens of developing nations, are often forced to suffer longer in poverty with its attendant high rates of malnutrition, disease, and mortality; as a consequence, they are often the most injured by such misguided, though well-intended, policies.

Our Beliefs

Our common Judeo-Christian heritage teaches that the following theological and anthropological principles are the foundation of environmental stewardship:

   ✝ God, the Creator of all things, rules over all and deserves our worship and adoration.

✢ The earth, and with it all the cosmos, reveals its Creator's wisdom and is sustained and governed by His power and loving-kindness.

✢ Men and women were created in the image of God, given a privileged place among creatures, and commanded to exercise stewardship over the earth. Human persons are moral agents for whom freedom is an essential condition of responsible action. Sound environmental stewardship must attend both to the demands of human well being and to a divine call for human beings to exercise caring dominion over the earth. It affirms that human well being and the integrity of creation are not only compatible but also dynamically interdependent realities.

✢ God's Law-summarized in the Decalogue and the two Great Commandments (to love God and neighbor), which are written on the human heart, thus revealing His own righteous character to the human person-represents God's design for shalom, or peace, and is the supreme rule of all conduct, for which personal or social prejudices must not be substituted.

✢ By disobeying God's Law, humankind brought on itself moral and physical corruption as well as divine condemnation in the form of a curse on the earth. Since the fall into sin people have often ignored their Creator, harmed their neighbors, and defiled the good creation.

✢ God in His mercy has not abandoned sinful people or the created order but has acted throughout history to restore men and women to fellowship with Him and through their stewardship to enhance the beauty and fertility of the earth.

✢ Human beings are called to be fruitful, to bring forth good things from the earth, to join with God in making provision for our temporal well being, and to enhance the beauty and fruitfulness of the rest of the earth. Our call to fruitfulness, therefore, is not contrary to but mutually complementary with our call to steward God's gifts. This call implies a serious commitment to fostering the intellectual, moral, and religious habits and practices needed for free economies and genuine care for the environment.

Our Aspirations

In light of these beliefs and concerns, we declare the following principled aspirations:

We aspire to a world in which human beings care wisely and humbly for all creatures, first and foremost for their fellow human beings, recognizing their proper place in the created order.

We aspire to a world in which objective moral principles—not personal prejudices—guide moral action.

We aspire to a world in which right reason (including sound theology and the careful use of scientific methods) guides the stewardship of human and ecological relationships.

We aspire to a world in which liberty as a condition of moral action is preferred over government-initiated management of the environment as a means to common goals.

We aspire to a world in which the relationships between stewardship and private property are fully appreciated, allowing people's natural incentive to care for their own property to reduce the need for collective ownership and control of resources and enterprises, and in which collective action, when deemed necessary, takes place at the most local level possible.

We aspire to a world in which widespread economic freedom—which is integral to private, market economies—makes sound ecological stewardship available to ever greater numbers.

We aspire to a world in which advancements in agriculture, industry, and commerce not only minimize pollution and transform most waste products into efficiently used resources but also improve the material conditions of life for people everywhere.

Part 2

A Comprehensive Torah-Based Approach to the Environment

Introduction

Young children often develop irrational fears of the world and find themselves haunted at night by a phantom menace until maturity—or a creative adult—successfully wipes away the tears. There's a story of a father who would regularly be awakened by his son's recurring nightmare, which was provoked by the boy's daily encounters with an overly affectionate dog. Several nights a week the man would rush into his son's room to calm a wild-eyed little boy with a racing pulse. There, the father would sit upon his son's bed while the boy pointed out half a dozen dogs sitting on the carpet waiting to munch on his toes. The young boy would sit in his father's arms trembling, while his father futilely explained to him that there was no pack of dogs at all. After several weeks interrupted nights had reduced the man to a mere shadow of his usual robust self, he knew something drastic needed to be done.

The next night when he awoke to his son's scream of terror, the man strolled calmly into his son's room and began rounding up the dogs. It took him no more than half a minute or so of arm waving and hissing to chase the six canines out of the room. The man was rewarded with a sleepy smile and a "Thank you, Daddy," as he staggered back to bed. After two more nights of being chased out of the room, the dogs never returned. In the son's somnolent state, those dogs were a real problem. Trying to persuade him that the dogs did not exist merely frustrated the boy. He felt stuck with a handicapped parent who foolishly responded to dangerous dogs with mere words. The boy's father had to enter his son's frame of reference and see the dogs in order to get rid of them, and ease his son.

What we have come to refer to as the environmental issue also possesses two distinct frameworks of reality. According to one of these views, there is no imminent peril that threatens to destroy us, just as there were really no dogs attacking the boy who lay safely in his bed. According to the other view, however, the problem is real, terrifying, and seemingly intractable. According to this view, the world's original condition of natural perfection is being irreparably jeopardized by human activity. Currently, with many persuaded of imminent peril, the panic is spreading, and a portion of the population is horrified by "nocturnal dogs" that come in the form of "threats to the environment."

This is not to say that there is no environmental problem. We are not dealing with an unhappy child and imaginary dogs. It *is* to say, however, that the real problem may have more to do with beliefs and convictions than with objective and quantifiable peril. This in no way simplifies the problem. Just as in the case with the small child, it is usually necessary to enter the framework in which the problem exists before one can effectively attempt a solution. In the bright light of sunrise, that little boy laughed at the nighttime intruders. At the time of the crisis, however, help came only from someone within the framework of his reality. If someone really believes the dogs are there, the problem is not the dogs but the belief.

If we believe and are convinced that no more important value exists, for example, than prolonging life span, we would be justified in prohibiting all activities that could abbreviate national life span averages. But as humans, we have always demonstrated that we are often motivated by other conflicting values. Soldiers often perform heroic acts that shorten their own lives. Many individuals choose to smoke, skydive, or climb mountains because of what these activities contribute to their lives, and they do so in full knowledge of the possibility that they may be shortening their lives. Environmentalism, especially in its more radical and virulent forms, frequently places the preservation of nature in the forefront of moral consciousness, above and beyond other values with which it may well be in conflict. In so doing, any calculation of relative benefits may be censured. We might also be making *facts* irrelevant to *judgment*.

People seldom argue passionately over facts. We tend to dismiss as foolish people those who argue over facts that are either known or easily discoverable. People might well debate which is the most beautiful mountain in the world, but now that technology permits us to take accurate measurements, they will never debate which is the highest. The purpose of the Torah, according to traditional Judaism, is to help us establish the correct beliefs with their profound ramifications, rather than to impart mere facts. Well-established scientific methods, on the other hand, provide the legitimate venue for resolving matters of fact.

Thus, the real environmental problem may well be the very belief that there exists a terrifying problem rather than any problem in itself. At the very least, it is a problem that is enormously exacerbated by certain beliefs that can stand in the way of a genuine commitment to stewardship of all God's creation.

We shall examine further in this essay the modern phenomenon known as environmentalism; we will look at the Torah's understanding of the "middle path" and how it relates to morality and human population; next, we will review the Jewish understanding of the right relationship between the human person and nature, especially as this relates to work and the creative spirit; and, finally, we will close with a discussion of the Torah's view of property, pollution, and the law.

I. Human Population and Achieving the Middle Path

Every year governments and prominent industrialists dedicate enormous sums of money to population reduction programs conducted by a variety of agencies ranging from Planned Parenthood to the United Nations Population Fund (unfpa). After all, the argument goes, it is obvious that there must exist some maximum number of people who can survive on "spaceship earth." We may not yet know what that number is, but that does not mean it does not exist. There must be some world population figure beyond which people will no longer have adequate food or enough resources to survive. And even if this turns out to be untrue, there surely must be some figure above which there simply will no longer be space for additional people to live. Granted, this number would be quite large, but as long as we concede that the annual growth in world population takes us inexorably closer, why not start doing something about it right away?

So what if all of America's population could comfortably live in the small part of California between Los Angeles and the Mexican border? All this means is that doom is not imminent in America. Clearly, in the far more crowded conditions of Africa or Asia, the argument continues, responsible leadership should demand immediate action. Not only is the welfare of entire nations threatened by unrestrained population growth, but so is the living standard of families within those nations. Too many children impose economic hardship on families who are discouraged from using "family planning" techniques by ignorance or religious taboo. These families require larger homes, use more water and heating resources, and shrink available "green space" within cities.

The argument appears formidable, and indeed it is. It is neither effective nor true merely to insist that people always find a timely and appropriate solution to their problems. Sometimes we do, but occasionally we do not. Against Thomas Malthus's stern warnings of two hundred years ago, we did find answers. New machines that made fabric plentifully and inexpensively could clothe those whom Malthus anticipated would be cold. Agricultural advances made food available for those whom he predicted would starve. For some problems, we never did find an answer. Some of the costliest wars of the twentieth century, for example, could have been avoided had we found a timely solution.

The Torah stresses a golden mean in problem solving. The great transmitter of Torah thought, Moses Maimonides, discusses how to achieve this "middle path," as he calls it, in his magnum opus, the *Mishneh Torah*. Visualize the two extremes, he advises, and then seek the geometric midpoint. For instance, neither extreme sternness nor excessive indulgence is desirable as a full-time guide to life. The excessively stern person could never raise a child without injuring him or her physically, whereas the intensely indulgent person could never raise a child without injuring him or her spiritually. This person would never be capable of exerting discipline or administering the occasionally necessary punishment. However, the parents who guide themselves down the middle path will be able to reach into themselves for the reserves of both stern discipline as well as soft compassion, as the situation demands.

Similarly, there are two extremes of human behavior, neither of which serves well. One extreme occurs when we totally ignore the future while living hedonistically and indulgently for the present. Parents feel pangs of pain while watching a growing child live self-indulgently with no thought for the future. The alternative extreme is that we can suffer through a present of complete self-deprivation in order to save for the future. Many of us have known people who survived the Great Depression of the twentieth century. These persons frequently lived the rest of their lives in depression-like circumstances, even though they possessed financial reserves that made the self-deprivation unnecessary. The challenge facing the person wishing to live the good life is to find a more balanced approach. One of Judaism's great gifts to its adherents is a "manufacturer's guide" to how the human person can best attain this middle path. The Torah provides a roadmap to achieving balance—being neither a miser nor a spendthrift, being neither a libertine nor an ascetic. The middle path enables one to live each day to its maximum joy potential while also conserving resources for an unknown future.

The Torah's response to the population panic is consistent, teaching us first to identify the two extremes. One extreme is to invite government to impose draconian regulations and arduous restrictions upon us. This view insists that no sacrifice today is too great in the attempt to diminish tomorrow's threat, no matter that the precise nature and time frame of the threat remain unknown. The opposite view, in the words of Nobel Laureate Jan Tinbergen, maintains,

> Two things are unlimited: the number of generations we should feel responsible for, and our inventiveness. The first provides us with a challenge: to feed and provide for not only the present, but all future generations, from the Earth's finite flow of natural resources. The second, our inventiveness, may create ideas and policies that will contribute to meeting that challenge.

So we see that one extreme is to regard no sacrifice today as too much to impose upon ourselves to protect all future generations until the end of time. Had earlier generations followed this perverted logic, they might well have restricted the use of whale oil. One can imagine the decrees emanating from zealous eighteenth-century environmental activists, banning the use of oil lamps past nine o'clock at night to ensure that sufficient whale oil would remain to light the homes of the twenty-first century. In so doing, what they may well have effected is limiting the educational possibilities of the early scientists who studied and experimented late into the night to discover petroleum and its many uses.

The paradox revealed by the Torah is that far from solving any problem, following either extreme actually aggravates the underlying situation. This is one of the reasons that Judaism insists on a child being raised by both a man and a woman wedded into one. A healthy child needs to be raised with both the discipline and firmness that is the natural characteristic of the male as well as with the gentleness that comes so easily to the female. Guided only by the paramount principle of indulgence or by its counterpart, cruelty, raising a child will, in both cases, yield a monster. Only the balanced middle path offers any hope of raising a well-rounded person.

Similarly, we can either ignore the growth of the human population or we can impose limits on it. If we simply ignore the problem—insisting that there is no problem—we make the same mistake made by the father when telling his son that there were no wild dogs in his room. At best, ignoring population growth does no more than persuade the population-panic enthusiasts that we are blind. At worst, it may really blind us to what may turn out to be a valid concern. On the other hand, imposing oppressive regulations of either the criminal or the tax-policy variety or promoting an ethic designed to limit families to one or two children, for instance, will also aggravate the problem in a manner already conspicuous in India, Korea, and many other parts of Asia. One unintended consequence of the population policies that have already been in force in these countries for several decades is a severe imbalance in the sex ratio. Planners are already discussing the grim picture presented by the soon-to-arrive specter of several million Asian men unable to find wives.

Thus, whether we choose one extreme or the other, we will worsen the situation we are hoping to resolve. Is there a Torah approach to the so-called "population bomb"? Naturally, the proper approach is the balanced middle path. We should not ignore the problem, but neither should we precipitate chaos today in a foolhardy attempt to ward off a distant threat, one whose outlines are still dim and vague. What is this mysterious middle path? To discover it, we need to review our fundamental beliefs about whether a human being really is a consumer or a creator. If man is merely a consumer, then, obviously, the fewer, the better. If, however, man is a creator, then, equally obviously, the more, the merrier. And the answer is not "both." That would settle nothing. What we are asking is whether humans create more than they consume or consume more than they create. The Torah answers its own question: Humans can be either consumers or creators. This is quite a different answer from saying "both."

The Torah-true answer is that we can raise children to be either consumers or creators. If we raise them as if they were young animals, they will grow into animals—basically consumers who are able to work like horses, but never with the capacity to truly create. In order to achieve that ability in our children, we have to raise them in the image of the ultimate Creator. That means imparting to them a sense of limits, an awareness of what is right and what is wrong. Only animals have finite needs. Humans, touched as they are by the finger of the Infinite Divine, have infinite wants. Children have to be taught that every want will demand a choice and a sacrifice, and that each of us must responsibly steward what we have been given and what we have earned. Children deserve to know that while we relate to and sympathize with their feelings, we do not expect them to follow those feelings unthinkingly. We expect them to follow their head, not their heart. They should grow into the realization that the world is not necessarily a fair place, but that it does have rules. Knowing those rules is better than whining about fairness. Finally, they should know that life judges us by our performance, not our intentions. Children raised to live by these and other similarly true and enduring principles, are a pleasure to be around.

How exactly does raising the right kind of people help to solve the problem of too many people? The Talmud relates that during the pilgrimage festivals, the Jerusalem Temple was so crowded that people barely had room to stand. However, during the period of the service that called for worshippers to prostrate themselves upon their knees on the floor, there was mysteriously sufficient room. This is, indeed, a mysterious account since everyone knows that people on their knees require more floor space than people standing erect. During the part of the service when people were on their knees, conditions should have been more, not less, crowded than when the people were standing. The traditional explanation is that standing erect is a metaphor for a condition of arrogant self-absorption. Prostration is a metaphor for humility and awareness of others. Finally, the Temple itself is depicted in the Torah as an almost mathematical model of the world. It is not hard to grasp the truth of this message: If a population consists of humble people constantly aware of one another, it never feels crowded. However, if a population finds itself surrounded by even a few arrogant and self-centered individuals, conditions feel overcrowded. Overpopulation is not a question of numbers or objectively measurable figures such as people per square mile. Instead, it is a question of whether people feel oppressed by the overwhelming presence of others. This has more to do with standards of civility and behavior than with actual population numbers. Most of us would feel less pressured and more comfortable on the crowded streets of Hong Kong or Tokyo than we would on a lonely urban alley in New York City. What we really have is not a population problem, but a perception of a population problem—a problem that results not simply from too many people, but from too many people arrogantly and thoughtlessly impressing their presence upon others. Rather than reducing the number of people, we need to reduce the incidence of selfish behavior that oppresses others and to increase the amount of creative behavior that meets others' needs.

This may seem an inadequately poetic prescription for a pressing and prosaic problem, but it is really all we have. To seek one extreme, by doing nothing and merely watching as selfish and coarse children are born and raised to crowd a culture, is foolish. Naturally, we will all come to feel that there are too many people. We have to do something. However, seeking the opposite extreme of encouraging fewer people while ignoring the behavior of those people is equally foolish. It should be noted that this is true as long as the threat of overpopulation is vague and distant. All that is left for us to do is to focus on inculcating into our culture those values that would diminish the perception of overcrowding and also increase the contribution made by each member. This would not only reduce the clamor for population control but would also make for much more tranquility and considerably more prosperity for all of us.

II. The Right Relationships Among God, Man, and Nature

In the prevailing climate of the environmental debate, it is necessary to state categorically at the outset that the Torah unhesitatingly prohibits cruelty to animals. This is not because animals also have rights; it is because only human beings have obligations. In the Torah's depiction of moral reality, nobody has rights—only obligations. Naturally, if everybody discharges their obligations, we all end up enjoying those things we vainly attempted to obtain by claiming them as our rights.

The animal rights movement can best be understood by viewing it as an attempt to undo the opening chapters of the biblical Book of Genesis. The Torah and its accompanying oral transmissions insist that Genesis describes more the beliefs underlying Creation than its facts. This is to say that the Bible's central premise is that humans and animals are qualitatively different, a contention violently opposed by the animal rights movement. After all, a woman wearing a fur coat is offensive only if she is nothing more than an animal as well—a very intelligent and well-evolved animal to be sure, but an animal nonetheless. And wearing one's cousin's skin over your shoulder is simply barbaric. Animal rights advocates insist that we are all animals, and no animal should have any special, species-specific rights that all other animals do not also enjoy.

The Bible teaches that the human person is the apex of God's creation and that all creation is there for the human person to develop and use as a responsible steward. The principle at work here is, of course, precisely the same biblical principle that prohibits self-maiming, destroying a rented apartment, or even having an abortion. This is to say that tenants do not have the same rights as owners. We, as humans, do not own the world, our bodies, or the habitations we rent. Thus, we may improve them but not destroy them. According to the Torah, not only do women not have the right to do with their bodies as they wish, but neither do men. Our bodies are given to us by a gracious and generous God so that we may occupy them for a certain period of time. During that time they are to be treated with the same deference that a tenant should employ in caring for his rented premises. Similarly, we humans are granted use of the world and all it contains. We may hunt animals for food or clothing, build homes out of the wood we cut from trees, and mine the earth to extract the minerals it holds. However, we may not wantonly destroy anything at all.

Some of the areas in which animal rights activists have sought to infringe upon the rights of their fellow humans include efforts to curtail important, life-saving medical research; outlaw clothing made from animals; ban circuses; and damage the fur, meat, and poultry industries, sometimes through violence and intimidation. It is important to understand that they have taken these actions, not as the result of measurable data, but as the consequence of their own belief system. There exist two separate and utterly incompatible belief systems regarding animals. One of these doctrines stems from the belief that God created the world and all it contains, and, when done, created man as his deputy to further his work. The other doctrine stems from the belief that by a lengthy and unaided materialistic process, primitive protoplasm evolved into Bach and Beethoven.

According to the latter view, the human person is nothing more than a sophisticated animal. To devotees of this secularist faith, animal rights should indeed become the sacrament of secularism. There is no way to satisfy adequately both sides of the animal rights debate. By their very name, activists betray their agenda. By aggressive evangelism, they intend to promote and advocate the belief that no qualitative difference exists between humans and animals. Needless to say, by encouraging the oppressive human behavior mentioned above, this belief adds fuel to those who promote the population panic.

It is chiefly because of the absence of any prevailing moral counterforce that animal rights activists manage so easily to infuse their faith into the general culture. The Torah depicts the entire account of the serpent enticing Adam and Eve as a tug-of-war between man's divine nature and his animalistic inclinations. Classical Judaism recognizes a sort of spiritual gravity that inclines humans to view themselves as animals. As animals, we would have few, if any, moral obligations; we would be free to act in accordance with whatever we believe are our instincts; and we could follow our hearts instead of our heads.

As the poet John Milton describes so faithfully in *Paradise Lost,* Adam and Eve do succumb to their animalistic inclinations, but finally atone and recover their place as God's special children, created in his image and charged with the task of improving the world by populating it and conquering nature. The Hebrew for conquering, *koveish,* clearly distinguishes between annihilating and conquering. The former is a verb for utterly destroying one's enemy. The latter refers to leaving one's enemy's resources and abilities intact, or even enhancing them, but redirecting them for one's own end. That is what we are told to do with the resources of the natural world. We may not destroy, but we may use them in every possible beneficial manner. Animals are part of the natural world, and their purpose is strictly in the context of human life. One reason that sacrificial rites played such a vital role in the daily services of the Jerusalem Temple was to drive home the point to the ancient Israelites that killing animals in the service of God, and for the purpose of his people, was morally permissible.

A religious Jew may choose to restrict his diet to vegetables during the week, but come Saturday and most holidays, he is to eat some meat as a religious obligation. The reason for this is that God created a world of hierarchy. Minerals are consumed by a higher life form, namely plants. Animals survive by consuming plants, while the highest life form of all, humans, eat animals. It is interesting to note that those animals permissible to Jews as food are animals that eat only plants. In other words, those animals that violate the hierarchical order, such as wolves and bears, may not be eaten by Jews. Now, for a Jew to attempt to improve on God's definition of morality by refraining from eating any meat on moral grounds is another way of announcing that one is nothing more than an animal oneself. Animals are supposed to eat only plant life. Thus, a Jew who eats only vegetables is announcing himself to be a very good animal. Once each week, God demands of his people that they leave the moral refuge of vegetarianism. We are then forced to confront the reality that an animal died to provide our meal. That places an obligation upon us to be worthy of the sacrifice. Now, for an animal to die for no reason other than to provide meat for another animal is less than ideal. Thus, the plundering animal is regarded as *non-kosher,* or not fully worthy of being eaten by Jews. However, the Jew who eats meat on a regular basis knows that he must conduct himself in a manner that makes his food's sacrifice morally justified. He is obligated to be a human, not merely another animal.

While always prohibiting cruelty or wanton destruction, Judaism abhors the entire notion of animal rights since it violates the very foundation of biblical belief in God's sovereignty and God's role as ultimate arbiter of moral right. Judaism and secularism are fundamentally incompatible, and the doctrine of animal rights is a doctrine of secularism.

III. The Spiritual Nature of Human Work

The religious Jew has much appreciation for the beauty of nature. We are filled with gratitude for these natural treats to our senses that are also natural resources vital to the human race. In fact, a collection of benedictions is part of every religious child's early-learned faith arsenal. From the earliest age, Jewish children smilingly utter the benediction for a rainbow upon seeing this arc in the heavens. When seeing a beautiful tree, the ocean, hearing thunder, and for many other manifestations of God's world, we say a fervent "thank you."

But factories and skyscrapers also reflect Jewish values. A factory speaks of the human yearning to emulate God's power to create. A city speaks of humans living together in peace and harmony as instructed by their Father in heaven. For this reason, the Temple was to be constructed in the heart of

Judaism's quintessential city, Jerusalem, rather than in a remote corner of unspoiled countryside. While forests and swamps are certainly recognized to be part of God's creation, merely leaving them in their original and pristine condition is ignoring God's directive to harness the forces of nature for the benefit of the human race. We are to leave our imprint upon the world in a way that improves what we found. The metaphor is the gracious landlord who allows rent-free tenancy in a not yet fully completed home, asking only that its tenants constantly work to improve its condition. Leaving it as we found it is poor repayment for the generosity.

The general hostility toward industrial development that is often evidenced by environmental activists is frequently rooted in a pantheistic opposition to the God of Abraham, Isaac, and Jacob, and is as old as the Tower of Babel. Judaism takes note of how industrial development tends toward the spiritual and away from the merely material. In our own times, this is quite clear as we see development lead societies past the manufacture of steel and large machinery to the creation of data and knowledge. One hundred years ago, Americans were building ships and railway locomotives. Today, that work is often being done by more recently emerging economies, while we have marched on to produce products whose value per unit of weight vastly exceeds anything that was produced by our old heavy-industry economy. Judaism views this as a movement toward human recognition of the primacy of the spiritual over the material. It is no coincidence that this tendency for society to move toward the spiritual also brings along with it less disruption of nature. Instead of imposing barriers to industrialization upon the developing world, we would be better served to assist developing nations in moving through this early phase of growth. In this fashion, each part of the world can make its own decisions and judgments about how it will balance its own needs. There are parts of the world—and will probably always be parts of the world—where immediate access to food and shelter trumps all other concerns. Those of us in the developed world may not want a rubber-tire factory next door. However, if we lived near Cairo and presently were neighbors to the world's biggest garbage dump, which is populated by ghostly skeletons rummaging through the filth to find food for another day's existence, we may welcome the arrival of a tire plant to displace the garbage dump. Judaism has great faith in the ability of ordinary human beings to make their own decisions and to find ways to overcome tragic circumstances.

This faith comes from another religious conviction not shared by many environmentalists. Again, if we are nothing but sophisticated animals, it is only right that important decisions should be made for us by an elite group of people playing the roles of zookeeper or farmer. In this view of reality, we are not capable of determining for ourselves just how much prosperity we are willing to sacrifice to halt development. Since nature is the ultimate good, our zookeepers will determine that no burden is too heavy for us to shoulder in service to our god of nature. Judaism insists that we are exalted creatures built in the image of our Creator and equipped with almost god-like powers to create. Thus, Judaism opposes attempts to deprive humans from making their own personal choices; we each have the freedom and the responsibility to order our own behavior toward God's law. Naturally, Judaism also does not protect us from our own poor choices. Part of moral growth is living with the consequences of bad decisions. Part of Judaism's preoccupation with an oral transmission is the ongoing accumulation of experience that validates the Torah's laws.

The basic Jewish principle of balance and middle path also conflicts with the contemporary environmental doctrine that preserving each spotted owl and each kangaroo rat is more important than any costs borne by humans and any sacrifices made by people. Judaism would never countenance loggers suffering the indignity of joblessness in order not to disturb the nesting habitat of the owl. When homes for people become dramatically overpriced because of the regulatory costs of providing for the habitat of the kangaroo rat, Jewish tradition also must object. People need not justify their needs or desires to nature. They are warned only against destroying things for no good purpose.

The view being presented here is occasionally made less palatable by the admittedly immoral practices of some of the participants in our economy. When a large and powerful corporation inflicts measurable damage upon its neighbors, for example, and then takes refuge in legal tactics, a wellspring of local frustration understandably bubbles up. Morality cannot allow people to evade responsibility by hiding behind the corporate veil. The corporation is nothing more than a vehicle for human cooperation. By surrounding a disparate group of people with a culture, an ethos, and an entire system, the corporation allows individuals who otherwise might have to be subsistence farmers to cooperate with one another in a larger and more lucrative enterprise. This cooperation allows for the provision of goods or services to their neighbors in such a manner as to allow them all to derive desirable income from the venture. Nonetheless, a corporation possesses no right to inflict upon its neighbors damage that its employees, managers, or shareholders would be prohibited from inflicting individually.

We see, therefore, that Judaism views development as people following their Creator's mandate to be fruitful, to multiply, and to conquer the earth. Instead of maintaining a sentimental and false image of nature, we religious Jews understand that nature is harsh and unforgiving. We understand that since the expulsion from the Garden of Eden, the struggle imposed upon us by God is to extract a living from an often reluctant earth. We must do so without laying claim to the benefits of another's labor and without recourse to dishonesty or theft. Our task is, in essence, to subdue nature and redirect it for holy purposes. Even the traditional Jewish practice of circumcision speaks to this godly mandate. The world I gave you is not perfect, says the Almighty. Even your own bodies await your finishing touch. Even more so, we are told, the entire earth awaits your finishing touch. Your labor is welcome, and its results are pleasing to me, says the Lord. For this reason, Judaism is prouder of man's skyscrapers than of God's swamps, and prouder of man's factories than of God's forests.

IV. Pollution, Property, and the Law

There is little question that Judaism and its comprehensive legal system consider pollution to be a serious offense. Numerous examples of how one citizen can harm another by various forms of pollution are cited in the Talmud. However, these examples are always civil cases brought by one individual seeking damages against another. Conspicuously missing is the notion of government initiating action against citizens. One explanation for this is the Torah's strong enthusiasm for private and relatively free transactions between individuals. In Judaism, ecclesiastical authority is also civil authority. Thus, in an ultimate sense, our "central government" is God and the moral law. The Jewish king is instructed to write his own copy of the Torah personally, meticulously copying it from the official texts. He is further instructed to always carry it with him to indicate that he, too, is subservient to its rules and laws. The prototypical Jewish model of a king is King David, whose closeness to God resulted in his writing the Book of Psalms. He also worked closely with the high priest and the Sanhedrin, a supreme court made up of seventy-two rabbis. This model of a religious scholar-king is hardly the picture of a strongly centralized government.

There is thus great dependency upon the local court of law known as the *Beth Din,* or house of law. One enormous benefit derived from retaining a strong local flavor to law is that there is far less likelihood of cases arising in which an individual is charged with harming all of nature, all of the world, or all of the air and water. Cases brought before the *Beth Din* must be brought by the individual being harmed. Certain problems are simply too large for mere mortals to solve and are regarded as being God's problems; we turn to him in perfect faith to solve them. It would be considered an act of spiritual arrogance to usurp responsibility for problems of a cosmic scale. Is this the same as doing absolutely nothing about real pollution problems? No, not at all. By far, the majority of real pollution problems do indeed have local parties as litigants, and they are subject to local solutions and are addressed in Jewish law.

Jewish thought traditionally views these problems through the lens of religious faith. There is no certain way to answer the question of what will be the end of the human story. However, the question clearly has only two possible answers: either oblivion or deliverance. Perhaps we are all ultimately doomed by carbon monoxide, global warming, a rising tide of disposable diapers, melting polar ice caps, ultra-violet radiation penetrating a hole in the ozone layer, a rogue meteorite, nuclear winter, some combination of all the above, or some entirely new and unknown threat. The details are not important, but the conclusion is. One way or another, humanity is doomed. The only alternative is that through some grand program of divine redemption, all of humanity will be delivered into a new and better tomorrow.

There is no way to predict which will ultimately come to pass. We can, however, solve those problems that affect some real individual persons here and now. Is someone being harmed by the polluted rainwater run-off from his neighbor's industrial enterprise? Is someone's property value being adversely impacted by bad smells or noxious fumes (air pollution) emanating from his neighbor's activities? Is a landowner along a river bank polluting the water, thus harming those downstream? All these are examples of legitimate pollution problems addressed by Torah law.

There is, however, little Torah justification for exploiting human fears about the future to expand the role of government. Judaism would clearly resist the notion that we must tackle those problems that are too big for any human to solve by making a government big enough to try to solve them. Consider the prophetic warnings about the dreadful consequences of appointing a king. Absolutely no Torah precedent or theological justification exists for government imposing restrictions upon individuals for the benefit of "nature" or "the environment." Not only is this not an explicitly Jewish religious imperative, but the exercise of government authority for possibly dubious ends represents a clear rejection of traditional Judaism, which has always stood rock-solid against allying itself with the changing fads and fascinations of the moment. Orthodox Judaism criticizes those who attempt to keep Judaism up-to-date by importing the doctrines and movements of secularism. A few generations ago, Russian rabbis castigated those well-intentioned Jews who established Jewish communist groups with the goal of retaining the involvement of young people in Judaism. Today, similar misguided efforts establish Jewish branches of feminism, homosexuality, and radical environmentalism for the purpose of "keeping Judaism relevant." The core of Judaism has always been relevant precisely because of its commitment to unchanging values and its indifference to the philosophical fads of the day. According to Maimonides, the eleventh-century Jewish sage, "It is clear and explicit in the Torah that it is God's commandment, remaining forever without change, addition, or diminishment, that we are commanded to fulfill all the Torah's directives forever." Thus, large-scale fears such as the threat of world annihilation are best responded to by the Jew with faith that God will solve them. Meanwhile, we should each concern ourselves with acting in accordance with the covenantal rules. We may not damage our neighbor's property, but neither does our neighbor have the right to interfere with our activities of fishing, hunting, manufacture, mining, or agriculture, if these activities do not directly harm him or his property.

Judaism also resists the government taking control over more and more of a society because of its commitment to people owning property rather than a society owning property. One of the very few exceptions to this rule was the Jerusalem Temple that was, of course, owned by no individual Jew. Otherwise, much religious emphasis is placed upon people owning property, and much care is exercised to protect people from threats to that ownership.

It should be understood that the Jewish emphasis on private property is a religious manifestation of a people's relationship with their God and the moral law. Along with so many other aspects of Jewish life, this one also is intended to affirm the Genesis account of Creation, whose central thesis is that we humans are qualitatively different from animals. No animal owns property. To be sure, many animals exhibit a territorial imperative. For instance, lions and elephants both mark their territories to let others know they claim dominance over that area. However, this is not ownership. Lions do not object to elephants in their territory, and they depend on deer ignoring those border markings. If all animals respected lions' "ownership" of an area and kept out, lunch with the lions would be an unusual event.

The Book of Genesis, however, details the mechanism by which humans can own land. Abraham's purchase of a burial site for Sarah is presented in such detail precisely to familiarize Abraham's descendants with the methodology by which humans can own land. This methodology turned out to be a startlingly novel concept, not only to Ephron and the men of Chet, but also to far more recent nations and races that knew nothing of land ownership by people. Yet Judaism is clear that God's plan for humanity calls for people to own land. This is partially on account of God's desire for us to recognize ourselves to be different creatures from animals, and partially on account of God's desire that we live among one another and interact with one another. Economic interaction and its attendant rewards of wealth are part of God's plan to ensure that the children of God do constantly interact with one another for mutual benefit. Land ownership helps to ensure this dynamic.

It is worthwhile to note that God promises Israel very specific benefits to following the covenant, and these promises are very much benefits of this world. God ensures rain in its time, bountiful crops, happy homes, well-behaved children, and wealth—wealth like that which the faithful Job lost and then recovered. God safely makes these promises, as it were, because the covenant is more than mere ritual. It is far more than prayer and good deeds. Major parts of the covenant are focused on how to organize human society and its economic interactions. There are far more rules about human economic interaction in the Bible than about all the prayer and the dietary rules combined. These rules promote human interaction, mutual dependency, and wealth creation. Besides prohibiting each and every one of us from destroying things purposelessly, these rules further God's plans for humanity.

Conclusion: Theocentrism or Secularism?

Perhaps the most fundamental question that shapes almost every facet of the environmental debate is how humans arrived on this planet. There are clearly only two possible answers to this question. Either a benevolent and loving God created us in his image and placed us here, or, alternatively, we are here as a result of an interminably long process of unaided materialistic evolution that converted primitive protoplasm into each of us. Needless to say, the approach that claims that God used evolution to place us here merely attempts an answer to the question. Of course God could have used evolution. That is not the issue. The issue is only whether we were put here by a creator, or whether we arrived here by a random and unaided materialistic process.

If it is the former, then the Creator's views and wishes as expressed in his instruction manual on life, the Torah, need to be taken into account as we organize ourselves. If it is the latter, then there is no Creator and no instruction manual, and we are free—no, obliged—to follow our own best instincts. And the harrowing aspect of all this is that it cannot be settled in time to determine the best course of action. We have no recourse but to believe one way or the other. This is only a matter of belief, not facts. If it were a matter of fact, there would likely be no believers either in God or in materialistic evolution left, just as there are no true believers in the flat-earth theory or in old theories about heat being an extruded liquid. Facts tend to sort themselves out. Beliefs can be debated forever. Yet most of the genuinely meaningful decisions we make in life depend on beliefs, not facts. When people get married, it is with the belief that they are acting wisely and that they will live happily ever after. They act on the basis of belief rather than on the basis of any real and reliable facts.

Similarly, most of us lack the ability to determine, beyond any doubt, the facts concerning human arrival on this planet. Rather, we tend to intuitively recognize the subtle social consequences of either belief, and we then adopt the one that offers our souls the least dissonance. Those of us comfortable with the implications of divine rules and laws feel comfortable with God having put us here. Those of us committed to a life with no externally imposed rules and laws feel more comfortable with a belief that rules out a Creator. Not surprisingly, all our presumptions about environmentalism fall into place according to this simple schematic.

If there is no God, then indeed there is nobody to take care of future generations—nobody to care for cosmic threats to earth, nobody to solve the really big problems that will possibly face the distant future. It then becomes not only wise but also noble and moral to make that selfless worry for the future our own concern. If there is no God, then we humans are no better than any animal, and we only practice an evil form of "speciesism" by eating, using, being entertained by, or riding on animals. If there is no God, then any human conceit that we may change the face of the planet in a way that no animal would dream of doing, is just that—a conceit.

If, on the other hand, there is a God, then everything changes. If there is a God who has created us, then each and every human person has infinite value, and none can be sacrificed for the sake of nature or some abstract cause. It is God's definition of morality that we must follow. Recognizing that life is indescribably complex, Judaism disdains moral governance by aphorism. A Jewish judge is not someone who has exhibited compassion, intelligence, or popularity. Instead, an appointed arbiter of communal morals is someone who has been sufficiently familiarized with God's view of the extended order of human cooperation that we call society. This person would have done this by mastering not only the several hundred chapters of the Five Books of Moses, but also the thousands of pages of the Talmud and the thousands of responsa, which constitute the establishment of legal precedent during two millennia of Jewish jurisprudence. Spurning spurious simplicity, Jewish law even lacks a term for nature. While the Hebrew word *teva* does mean *nature*, it is not a word that can be found in the Torah, the Five Books of Moses. The omission is particularly noticeable in the first few chapters of Genesis, wherein God does not create nature. Instead, God creates each element separately. God creates mineral, vegetable, and animal, with all the subspecies and variations within that category. Traditional teaching insists that this understanding of Creation is to discourage worship of nature.

It is not possible to have it both ways. We must choose between two incompatible beliefs. One is the God-centered or theocentric view of reality to which each and every Jew is surely obliged to cling. The other, environmentalism, particularly in its more radical and virulent forms, is fundamentalist secularism. Those of us who consider ourselves persons of faith allow the environmental movement to set the terms of the debate at our own peril. The question is not how we should tackle and ultimately solve the problems about which environmentalists warn us. The question is how we should cope with more and more of our fellow citizens adopting a faith that inspires its believers to act in ways that sacrifice the multitude of human values to an environmental cause.

Clearly, to begin with, we need to demonstrate that we see the dogs in the dark room. We need to familiarize ourselves with the spurious science that produces terrifying scenarios on demand. But, in the final analysis, the child will be cured only when he no longer sees imaginary dogs, and when he walks confidently with his own dog at his side. The problem is not threats to the environment; it is really the threats to our souls. And as in countless earlier instances of history, imprudent beliefs can cause well-intentioned people to do terrible things.

Editorial Board

Part 3

The Catholic Church and Stewardship of Creation

Introduction

As Roman Catholics seeking to be faithful to the fullness of God's truth, we offer the following reflections in the hope that they will shed some much-needed light upon the environmental issues that are currently facing our world. We do not speak here as authoritative representatives of the Church's Magisterium, but only for ourselves as members of Christ's Mystical Body, reflecting upon environmental questions with the aid of Church teaching. This teaching derives its authenticity from its origin, which is Christ himself, and has been passed down to us by the Scriptures, Sacred Tradition, and the teaching office of the Magisterium. By the very nature of the Church's "catholicity," this teaching is intended to be universal in its scope, and, as such, has much to contribute to a proper understanding of environmental stewardship.

Because this teaching represents a two-thousand-year history of reasoned reflection upon divine revelation, it serves as an indispensable point of departure for establishing a deeper understanding of the created order, the nature and ontological value of God's creatures, and, in particular, humanity's value and place in that created order. An authentically Catholic understanding of the environment must be informed by a knowledge of these truths so that we can appropriately respond to environmental questions in a manner that respects the order that God has established. At the same time, a genuinely Catholic approach to environmental stewardship must constantly bring the moral authority of Church teaching to bear on all environmental questions. Thus, in addition to authentic scientific and reasoned analysis, even the most simple choices regarding the environment must be properly ordered to the truth about man and the world that is his home.

I. The Goodness and Order of God's Creation

If one takes the time to study the religious views of many ancient cultures outside the influence of revelation, one will notice how deeply our Western understanding of creation, God, and man has been shaped by the Judeo-Christian tradition. What ancient cultures provide for us are examples of the insufficiency of human reason in trying to penetrate the deepest mysteries of life. Though the religious views of ancient cultures varied, what we see, beginning principally with Abraham, is a radical departure from what we now refer to as paganism. Of the beliefs common among ancient peoples, a number of fundamental presuppositions seemed to figure prominently in their religious belief. For the sake of space, we will list them below:

Polytheism:

- Asserted the existence of many gods.
- Denied that human life has intrinsic value.
- Saw time as cyclical as opposed to linear.
- Lacked the understanding that objective moral norms emanate from the divine and are an essential component of proper worship.

Pantheism:

- Maintained that all of creation is divine.
- Saw time as cyclical as opposed to linear.
- Denied that God is a single, unchanging, perfect, transcendent, and necessary being who is totally above the created order.

Gnosticism:

- Maintained that creation developed out of a supernatural conflict between good and evil.
- Held that matter is evil, while the spirit is good.
- Sought to escape evil by transcending both time and matter.

As Catholics concerned about the environment, we believe it is important to establish the radical difference between a worldview informed by revelation and one that is not. One of the greatest concerns for the Church today in terms of environmental stewardship is the surprising emergence, among some religious and secular environmentalists, of what might be called "neo-paganism." Though the articulation of this new paganism may be far more nuanced and refined than that of ancient cultures, many of the fundamental philosophical and theological errors remain the same. The distinction between God and his creation has been blurred; man's place in the created order has been obscured, while creation is garnished with characteristics unique to persons alone. Consequently, much of the environmental agenda currently being advanced reflects an environmental ethic that stands in contradiction to the Church's doctrines of God and creation. It is our intention, therefore, to establish an environmental ethic that rests firmly upon the foundation of both sound reasoning and divine revelation.

At the very beginning of the Creed, the Catholic Church professes its belief in one God who created heaven and earth. That Creator, unlike those described in the pagan cosmologies of antiquity, is described as good—indeed, as the only good that is whole and perfect.[1] The opening pages of Scripture also repeatedly emphasize that the Creator looked upon his creation and "saw that it was good" (Gen. 1:4; 1:10; 1:12; 1:18; 1:21; 1:25). Of all his good creation, it is God's creation of mankind that completes the created order in such a way that he pronounces it to be "very good" (Gen. 1:31). The *Catechism of the Catholic Church* reinforces this fact: "*Man is the summit* of the Creator's work, as the inspired account expresses by clearly distinguishing the creation of man from that of the other creatures."[2] Human beings are described as part of that creation, as specially created in God's image and likeness, and as endowed with the unique powers of reason and will.

The order that is inscribed into the very fabric of creation reveals to us that not only is everything God created good, but also that creation itself reflects the grandeur of God. In the ancient tradition, the Church Fathers often spoke of nature and Scripture as two divine books. The first shows us some of God's attributes through traces and images of the Creator imprinted on material things. Among these attributes are his transcendence, sovereignty, and marvelous creative power that appear to us in the vast cosmos and the fertile earth with its wonderful assortment of creatures. Even some peoples prior to or outside the influence of revelation were moved by the wonder of the world to intuitions about its origin and how everything had been brought into being. The sheer variety of things led them to speculate about the plenitude of their source. The order and intelligibility they found everywhere seemed a trace of some divine reason or unitive principle operating in all creatures. The world's beauty and majesty spoke of some perfect spirit at work. Stars, seas, mountains, animals, and plants visibly pointed beyond themselves to some invisible reality hidden to mortal eyes.[3]

The biblical revelation deepened these intuitions still further, placing them on a firmer foundation, and encouraging believers to observe ever more closely the world God had made. The Wisdom Literature and the prophets testified to a profound experience of God's creative power and guidance over the world, and a sense of the awesome responsibility of the human creature. Or, as the Psalmist eloquently describes it:

> When I see the heavens, the work of your hands,
> the moon and the stars which you arranged,
> what is man that you should keep him in mind,
> mortal man that you care for him?
>
> Yet you have made him little less than a god,
> with glory and honor you crowned him,
> gave him power over the works of your hand,
> put all things under his feet.
>
> All of them, sheep and cattle,
> yes, even the savage beasts—
> birds of the air, and fish
> that make their way through the waters.
>
> (Ps. 8:3–8)

This vision combines the two basic dimensions of Scripture's view of creation: the glory and majesty we may contemplate in what God has made, and our surprising dignity as active stewards of the world, despite our mere creatureliness. This realization has echoed throughout Christian history. Saint Francis of Assisi best expressed the concrete implications of this insight in encouraging his followers to contemplate creation and to praise God "in all creatures and from all creatures."[4] It is no accident that the Franciscans, who loved and rejoiced in creation more than other religious orders, shaped individuals such as Roger Bacon. Bacon paid careful attention to nature and, as a consequence, figured prominently after the medieval period in the development of early experimental science.[5] Thus, in echoing a long-standing tradition, the Second Vatican Council declared that Scripture enables us to "recognize the inner nature, the value, and the ordering of the whole of creation to the praise of God."[6]

II. Christian Anthropology

As the summit of God's creation, man reflects the divine image in a most excellent way. Essential to this divine image is our capacity for reason, which enables us to know God, the world, and ourselves. We are also endowed with the powers of freedom and imagination that allow us to reflect upon our experiences, choose a course of action, and thus become cooperators in the opus of creation. It might be said that we ourselves are co-creators with God, and are consequently privileged in our ability to take what God has created and make new things, which creation, on its own, could not produce. This privilege bestows on us a dignity that surpasses other creatures precisely because we can participate spiritually in God's creativity in a manner that far exceeds the merely physical capabilities of other creatures. Furthermore, because the nature of human action is free and self-determining, these actions have moral value.

It follows, then, that with such capabilities, and by virtue of our dignity, God placed human beings in governance over his creation: "Let them have dominion over the fish of the sea, and over the birds of the air, and over the cattle, and over all the earth" (Gen. 1:26). This dominion was specified as a command to "till and keep" the garden, and was first manifested in the naming of the animals (Gen. 2:15–20). Since naming something is to know that thing's nature, we see the first manifestation of man's rational nature. Moreover, by the command of the Lord to till and keep the garden, we can assume that man was commanded to use his rationality in the governance of creation for the sake of bringing forth fruit from the earth. As is evidenced by man's original "nakedness," we can conclude that man's dominion over creation was intended to provide us with the means for sustaining and enhancing our existence. This stands in stark contrast to the animals and plants for which God's eternal law has provided physical attributes that sustain their existence. All of this occurred before the Fall, and it constitutes the originating Catholic vision of man's place within the created order.

Alongside these divinely and humanly acknowledged goods, revelation also warns, of course, about profound evils. The story of the Fall in the Book of Genesis explains why evil came into human hearts and societies. As the *Catechism of the Catholic Church* explains,

> man, tempted by the devil, let his trust in his Creator die in his heart and, abusing his freedom, disobeyed God's command.... In that sin, man *preferred* himself to God and by that very act scorned him. He chose himself over and against God, against the requirements of his creaturely status and therefore against his own good. Created in a state of holiness, man was destined to be fully "divinized" by God in glory. Seduced by the devil, he wanted to "be like God," but "without God, before God, and not in accordance with God."[7]

The original sin affected every dimension of human life. One of its results is that "visible creation has become alien and hostile to man."[8] Just as there have been, since Cain and Abel, unjust and immoral relations between persons, so, too, have actions been taken against creation. However, evil is not the dominant force of action in salvation history. God himself entered our world to redeem us through the Incarnation of Jesus Christ. By taking on human nature and restoring its original relationship to God, so began a process of recapitulation for us and the whole cosmos, which is "groaning in labor pains even until now" (Rom. 8:22). This has been accomplished so that we may hope that by Christ's ultimate action, "creation itself would be set free from slavery to corruption and share in the glorious freedom of the children of God" (Rom. 8:21).

We must be clear, therefore, about what dominion does and does not mean. While all things have been subordinated to human beings, we should rule over them as God himself does. This dominion does not grant to us the right to "lord over" creation in a manner incongruous with God's own manner of governance. Since the first moment of creation, God has provided for the needs of his creatures, and, likewise, has ordered all of creation to its perfection. Hence, man's dominion over creation must serve the good of human beings and all of creation as well. Thus, dominion requires responsible stewardship. Such stewardship must uphold the common good of humanity, while also respecting the end for which each creature was intended, and the means necessary to achieve that end. If man exercises dominion in a way that ultimately destroys nature's creative potential or denies the human family the fruits of creation, such action constitutes an offense against God's original plan for creation.

In thinking about our relationship with the environment, then, we must distinguish carefully between disordered human action, which harms creation and—by extension—human life and property, and responsible action, which the Creator intends for the good of the human family and creation. According to a pastoral statement by the United States Catholic Conference, "As faithful stewards, fullness of life comes from living responsibly within God's creation."[9] Nowhere does revelation suggest (as do some contemporary religious and secular environmentalists) that creation, undisturbed by human intervention, is the final order God intended. To the contrary, human beings, with all the glory and tragedy of which we are capable, are central actors in God's drama. Indeed, in the history of salvation, the human person and the natural world are never ascribed the same dignity. In the Sermon on the Mount, our Lord himself, while counseling his disciples not to be anxious and to trust in God's providence, assures them that God even takes care of the birds of the air, and adds, "Are not you of more value than they?" (Matt. 6:26).[10] The Scriptures frankly present an ordered hierarchy of being: God rules over all, and human beings serve as his stewards, exercising an instrumental dominion over everything, while also being accountable to him for our exalted position as the rulers of the earth.

Thus, we do rule—and are justified in subordinating and using nature for human purposes, so long as that governance is in accord with the truth about God's creation. As the United States Catholic Conference explains, humans bear "a unique responsibility under God: to safeguard the created world and by their creative labor even to enhance it."[11] Hence, the good steward does not allow the resources entrusted to him to lie fallow or to fail to produce their proper fruit. Nor does he destroy them irrevocably. Rather, he uses them, develops them, and, to the best of his ability, strives to realize their increase so that he may enjoy his livelihood and provide for the good of his family and his descendants.

Some would argue that if man refrains from exercising dominion over nature, nature would be better off. Yet the issue bearing the greatest importance is whether man would be better off. When man does not exercise dominion over nature, nature will exercise dominion over man and cause tremendous suffering for the human family. If we were to choose to refrain from exercising our dominion over creation, nature on its own would not necessarily produce the most advantageous outcomes for human well-being. Droughts occur, rivers flood, earthquakes strike, volcanoes erupt, fires start, and diseases infirm, causing harm to humans and other creatures of the earth as well. Why God in his providence allows such things to occur is a mystery bound up with the fact of original sin. The destructive consequences, however, are not so mysterious. Consequently, as rational beings, we have a primary responsibility to protect human life as a duty that acknowledges the dignity of the human person who is created in God's image. Our responsibility to care for the earth follows secondarily from this dignity, and, as such, presupposes it. We alone, of all God's earthly creatures, have the power, intelligence, and responsibility to help order the world in accord with divine providence and thus minimize the effects of natural evil.

III. The Lord of History

In part, man's prominence in creation derives from another dimension of reality revealed to us by God—that time does not exist as what might appear to be a never-ending circle of life. Time is not static or circular. We move through a history that had a beginning and will have an end. In fact, as Scripture indicates, the entire universe progresses along a linear trajectory that moves us closer and closer to some final end when the last chapter of history will be closed. What this might suggest to us is that creation is developing toward a final state of perfection. This is not to say that God's creation was imperfect at the beginning, but that creation is not finished and will achieve its final perfection as it progresses through stages of development until it reaches that end for which creation was intended.

Even recent science suggests that creation began with the "Big Bang," that the universe is perhaps fifteen billion years into its development, and that after billions more, our universe may simply dissipate. Even in secular terms, there is strong evidence for us to believe that nature and human civilization are intended to develop through time. Geology and biology have discovered that the very planet on which we exist is the product of long developmental processes. Almost all the elements on earth were manufactured in earlier generations of stars that burned out, exploded, and distributed their material into the universe. The great diversity of plant and animal species in our biosphere reflects the slow rise of more and more complex and varied organisms. In the human realm, the growth of civilization, with its patient advances in science, technology, social institutions, and religion, mirror, albeit at a quicker pace, what seems to be one of the central laws of creation—that greater and greater complexity or degrees of perfection take time. What should be noted, however, is just how much faster human civilization has developed by comparison to the rest of creation.

God has revealed that this historic character of creation is, for man, infused with religious significance. Scripture tells us that God, through his word, first created time and space, and then proceeded to make creatures to rule over these realms. Yet he placed man, at the climax, as ruler over the entire order (Gen. 1:3–26). Thus, God was the beginning, and the first cause, of creation, and the principle of authority from whom man receives his vocation to exercise his earthly dominion. Scripture also indicates that we are passing through time from our origin to some final end for which we were created—a final consummation in Christ (Rev. 21:5–6). Human history, in a sublime way, is unfolding and developing toward a final perfection in God himself. We also know from Saint Paul that Christ came "in the fullness of time" to redeem us (Gal. 4:4), and that he will come again at the end of history to judge the living and the dead (2 Pet. 3:1–10). In Christ, the fullness of God dwells, and in him all things find their fulfillment (Col. 1:15–20). Therefore, linear time and the development it entails are undeniable components of God's plan for us because they place a moral imperative upon man's temporal existence, and thus infuse human life with a more noble purpose. The fact that man was given dominion over the earth suggests that this final state of perfection, both for man and creation, will be achieved, in part, by the free employment of man's creative intelligence and labor upon the created order. In other words, God has commanded that we participate freely and intelligently in furthering the development of his creation. Because God has revealed to us that time has a beginning and an end, we must acknowledge the truth that human dominion over creation is infused with spiritual meaning and religious significance.

In contrast, many of the cultures outside the influence of divine revelation believed that time was cyclical. Such a view followed naturally from simply observing the life cycles of nature. Thus, ancient peoples often viewed creation as an eternal, self-perpetuating, self-sufficient, and self-contained reality. In short, creation was its own perfection. It was man alone who somehow existed outside that perfection and longed to embrace it. One can see a glimmer of truth in such a view. It certainly appears to be that way. Thus, the regularity of the seasons and the recurrence of certain life patterns were the most prominent features of existence.

God's revelation has elevated and perfected that view by situating the cycles of nature within the proper religious context of man's vocation. Thus, the wonderfully rhythmic cycles of creation, in addition to providing for God's creatures, are best understood and respected in reference to man's relationship to God. The cycles of nature and the regularity with which they present themselves reveal a principle of intelligence that draws man's attention to their source. The logic of creation, which can be discerned by man's rationality, helps us to transcend the merely material and guides us on our journey toward ultimate meaning. The ebb-and-flow, life-and-death cadence of nature is a sacrament of the living God that points to an absolute perfection that stands both above and under all things. Were it not for the splendor of creation, man would have nothing to contemplate, and thus, nothing to direct his glance toward God, nor any way of discerning the meaning of his existence.

Therefore, we will not fully understand God's revelation, human nature, or the integrity of creation if we limit ourselves to a cyclical view of time and nature. Just as the marvelous world in which we live came to its present state of development over time, so, too, must our religious and secular knowledge develop toward the fullest understanding of God's plan for us. Noah, Abraham, Moses, and Jesus represent crucial stages in this history of salvation, which unfolds in time. Thus, Sacred Tradition reveals to us that God is not only the Lord of creation, but the Lord of history as well.

Many persons who are concerned about our impact on the environment believe that linear thinking and action violate the Creator's intention of a permanent and stable natural order. However, this is a point where both revelation and man's achievements—particularly in the arena of good science—will correct this misperception. Nature and human society are dynamic systems that depend on both change and continuity for their existence. In any faithful reading of either the book of nature or Scripture, we can see that, despite our concerns about what the short-term environmental effects of development might be, we must continually raise our eyes to the larger perspectives of God's providence and his intentions for humanity. Environmental stewardship consists in discovering how to properly understand the relationship between cyclical processes and linear developments, present in both nature and human civilization, so that they coexist harmoniously, and direct us toward the ultimate good, which is God himself.

Basing our existence upon cycles alone would be a great limitation on human civilization. The great Christian theologian, Saint Augustine, who was familiar with the cyclical views of antiquity, saw in the Christian vision a great liberation of the human race. He states, "Let us therefore keep to the straight path, which is Christ, and with Him as our Guide and Savior, let us turn away in heart and mind from the unreal and futile cycles of the godless."[12] Elsewhere, Augustine speaks of God as marvelously creating, ordering, guiding, and arranging all things, "like the great melody of some ineffable composer."[13] As a reflection of this, the human person, who is made in the image and likeness of God, composes, writes, paints, dances, grows food, makes tools, manufactures, and brings forth many new things from the intelligibility inscribed into the very order of creation. Because man cannot create ex nihilo as God does, it is precisely the cycles and logic of nature that assist man in exercising his creative inclinations. In other words, while we depend upon the cyclical dimensions of nature for how we develop in our own earthly existence, we have within us the same creative thrust that set in motion the whole history of the universe. In effect, our creativity can bring nature to a higher degree of perfection. Thus, we are faithful to the potential God has placed within us when we affirm what is intrinsically good in nature by developing new and previously unrealized goods.

Interestingly, the Church acknowledges this truth through its liturgy. The very unfolding of the liturgical calendar itself and the celebration of liturgy reflect the times and seasons of the earth, celebrate the products of man's ingenuity, and then suffuse them with spiritual meaning. Every sacrament of the Church affirms God's blessing upon man's dominion over nature by the mere fact that God chose to communicate his grace to us not through the fruits of nature alone, but through those fruits that have been further developed by human intelligence. Thus, even in our worship, we affirm both the value of creation, and the value of man's creativity, which gradually brings all of creation closer to its final perfection.

IV. Human Labor and Human Progress

Not surprisingly, the imperative for human work to meet human needs and restore our fallen world, which is implied by the process of development, appears throughout Scripture. Adam and Eve were given stewardship over the Garden. Cain practiced agriculture, and Abel tended flocks, as did many of the Hebrew patriarchs; and David, the Lord's anointed, was a shepherd before he became king of Israel. In the New Testament, our Lord himself learned carpentry in Joseph's shop, which means that even the holy family had to support itself by humbly shaping wood into useful products. Several of the apostles earned their living as fishermen, and Saint Paul made tents so as not to be a burden to others. Even the holiest of Catholic sacraments, the Eucharist, makes use not of wheat and grapes, but of bread and wine, "which earth has given and human hands have made," thus reflecting the cooperation between God's grace and our labor in the work of salvation.

As necessitated by this tradition, the Church has subsequently placed great value on human labor as perhaps no other religion in history. Though this world is passing, for Christians the material world is not an illusion, as other religions have sometimes maintained. Thus, work and discovery are essential to God's plan for human fulfillment. The very work of salvation history itself has been unfolding here, in time, in space, and in the flesh. Likewise, the world for the Christian is not, as modern science suggests, a mere repository of raw materials and energy to be harnessed toward whatever purposes we feel inclined. To the greatest extent, the value of human labor finds its fulfillment in the discovery of those ways in which nature can be most responsibly and effectively placed at the service of the human family. This is the most authentic definition of human progress.

Christianity's affirmation of human progress is demonstrated throughout history. For example, out of loving attention to God's world, and the value placed on manual labor in the Benedictine monastic tradition, the

Western impulse toward material improvements, and the later development of science, partly find their origin.[14] Some of the greatest early modern scientists, such as Galileo and Newton, were Christians who thought of their work as faithfully discovering the nature of the world that God actually made. These observations of the physical world were, in part, made possible by medieval scholastic philosophy and its Aristotelian metaphysics. Had it not been for the work of people such as Saint Thomas Aquinas, the Scientific Revolution of the fifteenth century may have occurred much later or not at all. From the careful attention and desire to better the human condition that developed within the monastic tradition, eventually spreading into the universities throughout Europe, many valuable developments emerged, and human beings began more fully to understand and express their own God-given role in creation. In our own times, Pope John Paul II has stated,

> the earth, by reason of its fruitfulness and its capacity to satisfy human needs, is God's first gift for the sustenance of human life. But the earth does not yield its fruits without a particular human response to God's gift, that is to say, without working. It is through work that man, using his intelligence and exercising his freedom, succeeds in dominating the earth and making it a fitting home.[15]

However, a genuine concern has recently arisen that our very God-given capacities may, in fact, be endangering creation. Though man is the summit of creation, our burgeoning powers have made us acutely aware of the particular goodness, vulnerability, and interdependence of all creatures.[16] As Pope Paul VI observed, "The Christian must turn to these new perceptions in order to take on responsibility, together with the rest of men, for a destiny which from now on is shared by all."[17] This new situation, with its new perceptions, calls for a new ethical effort, and further broadening of the Catholic moral vision. The primary Catholic approach to the moral life focuses upon the development and habituation of virtue. Clearly, human action toward the environment must be guided by something more than utilitarian calculations and human wants, especially since those have been distorted by the Fall. How to apply a knowledge of virtue to environmental questions is complex and has only recently begun to be addressed. A full treatment cannot be offered here. However, a few brief suggestions are in order.

At the center of the moral life the Church identifies four cardinal virtues: prudence, temperance, fortitude, and justice. Briefly, these virtues are pivotal for establishing a norm of behavior for human action, and, for our purposes here, those actions which adversely affect the environment:

Prudence: As the mother of all virtue, prudence demands that we reflect deeply upon the highly complex particulars that are involved in environmental stewardship, along with those moral norms articulated in Church teaching. The most diligent application of prudence, however, will not solve all our dilemmas. Nonetheless, by prudently acknowledging the limits of our human knowledge and judgments, we will prevent ourselves from pursuing impossible utopias, and thus proceed cautiously toward the best possible solutions for both the good of the human family and the good of nature. Prudence necessitates humility in the face of complexity.

Temperance: As the virtue that restrains and directs our disordered appetites, temperance has obvious applications for environmental stewardship. It suggests that simplicity of life, self-discipline, and self-sacrifice, as Pope John Paul II has reminded us, "must inform everyday life."[18] Temperance is the virtue required for a proper ordering of consumption.

Fortitude: In earlier times, we needed great courage to face the challenges that the material world posed to our existence. Many of the discoveries that have benefited the human family required individuals to courageously discover the powers and potentials of nature. This tradition continues still, but with little regard for moral norms. While fortitude has often been of tremendous value, it requires that we avoid pursuing technologies that violate the natural law or could result in the mass destruction of nature and the human family.

Justice: As all people are impacted by ecological concerns, justice requires that each creature be given its due in accord with its own particular goodness. Consequently, where tradeoffs are necessary, human need must always be given priority. Wealthy societies are better able to absorb environmental costs, and, therefore, they should bear them; but they should also assist poorer nations in the process of economic development so as to help them secure their own dignity and will. In the long run, such efforts benefit both man and nature.

Some of these points will be touched on later in this essay. Nonetheless, it is clear that, for the Catholic tradition, virtue is an indispensable foundation to understanding how human beings are called upon by God to play their proper role in restoring and developing God's creation in accordance with his original plan.

V. Human Power and Nature's Ways: Some Prudential Considerations

The ongoing process of discovering potentiality in nature and choosing which portions of that potential to actualize, leads us into many complicated prudential judgments. The judgments we make here are not the only prudent conclusions from Catholic principles, but they seem to us the best reflection of sound theology and sound science.

For much of history, human interaction with the environment had few lasting effects. Nature was immensely powerful, compared with the limited capacities of mortal man. It is only the immense growth in human powers in the past few centuries that has made human activity a potential threat to the integrity of creation. Prior to that expansion, people in every part of the world over-fished, over-hunted, over-harvested, polluted, and, sometimes, harmed themselves and their fellow creatures in the process. However, the relative weakness of the human animal in the face of nature's immense power and fecundity made such damage local and transitory. Nature itself has produced much larger disruptions. During the last Ice Age, for instance, which ended only about fourteen thousand years ago, a large portion of the northern half of the globe was covered in ice thousands of feet thick. Forests were scraped clean from the land; few plants or animals survived. Yet the reproductive powers of life on this planet are such that the splendid northern forests we now enjoy reappeared in a relatively short time. Creation itself has a wide range of states as well as enormous regenerative powers when it is allowed to use them.

Some changes push the world into greater complexity and proliferating forms of life; others kill off species—and sometimes even whole ecosystems—without the slightest human intervention. What is often spoken of as the "balance of nature," therefore, is a dynamic balance. Nature changes all the time. In the past, for instance, the earth's climate naturally underwent fluctuations that were faster and larger than even the worst scenarios for manmade climate change. The course of rivers, as well as the locations of forests and of deserts, shifted without ceasing. These forces, which destroy only to create anew, seem to be part of the way that the Creator intended to bring about the intricate and varied forms of life we see around us today. If we think of the balance of nature as static, we will not only have a false impression of the world God has given to us, but we will work against the dynamism of nature and human nature, even as we seek to help both to flourish.

Nature is also sometimes described as a self-regulating system. Again, this is only partly true, and needs to be rightly understood; nature's way of self-regulation raises hard questions for responsible stewardship. Nature achieves balance when one portion takes advantage of opportunities presented by another portion. Big fish eat little fish. Weaker species reproduce in large numbers to offset the losses to predators. None of this, of course, is an ideal model for human individuals and societies to follow. We have concerns that no other earthly creature manifests. Very few of us, for instance, would wish to obliterate the natural beauty and varieties of plant and animal life around us, even if it would entail no physical harm to our own species. A healthy and beautiful environment is one of the goods man values, and, therefore, seeks to promote. By contrast, the HIV virus that causes AIDS does not care if it wipes out all potential animal hosts because the only thing it appears to know is how to reproduce to the limit of available biological niches. Other species seem to behave in essentially the same way.

Despite our natural affection for our fellow creatures on this planet, we need to see them as they are in themselves, and in terms of what they mean for human life. Elephants and tigers, for instance, are marvelous creatures that should be preserved; they tell us something irreplaceable about God's "infinite wisdom and goodness."[19] However, wild elephants and tigers have also been the bane of human existence, as have been viruses, mosquitoes, wolves, bears, sharks, and a menagerie of other creatures. To recognize this is not to license any and every human action over nature. Man's dominion over nature is "not absolute; it is limited by concern for the quality of life of his neighbor, including generations to come."[20] Still, persons who live in close contact with nature have a very different sense of its relative mix of threat and glory than do persons who observe beautiful rain forests or wild animals only at a safe distance through television, movies, or with the advantages of civilization to which to return. Nature contains many dangers for the human race, as well as much beauty and benefit. Some religious and secular environmentalists give the impression that it would be better for man and nature if we returned to some previous state, certainly before industrialization, and perhaps nearer to prehistoric conditions prior to settled agriculture. These aims are both wrong-headed and dangerous. Creation becomes benign for man and realizes potentialities built into it by the Creator to the degree that, through developing his own creative powers, man takes dominion over creation. Left on its own, nature is limited in what it can achieve by its own natural processes. Thus, nature would fail to release the potential God intended for it if not for the instrumentality of human creativity and labor. Furthermore, untamed nature would continue to inflict tremendous suffering on the human family.

VI. A Better Sense of Perspective

The modern concern about the environment, and the very development of the science of ecology, began in the middle of the nineteenth century when human power over creation began to expand rapidly. As we might expect, good and evil were inextricably mixed in this development. On the one hand, industrialization and modern agriculture have enabled more people to live—and live a more fully human life—than ever before. After a difficult transition period, for instance, manual laborers in advanced economies achieved a security and sense of dignity never before seen in any society. Advances in technology have made famine—which was a regular scourge to humanity around the globe before modern times—a thing of the past, except in places where political tyranny or turmoil prevent intelligent development. Advances in medicine have all but eliminated diseases such as smallpox, tuberculosis, and malaria, and have made formerly life-threatening maladies such as measles, mumps, and others, relatively minor nuisances. All of this was achieved by the slow and patient accumulation of human knowledge and the creation of free institutions that enabled the fruits of that knowledge to be shared by even larger numbers of people.

On the other hand, industrialization also had its negative effects. Early industrialization polluted cities, disrupted agricultural communities, and challenged modern nations to find ways to integrate growing urban masses. However, these were largely transitional problems. Today, it is precisely industrialization, new forms of agriculture, and other human advances that are making it possible for humans to increasingly live well and in proper relation to the earth. Even in difficult cases, such as the increase in greenhouse gases, we want to be wary of taking too narrow a view of the matter that neglects a broader perspective on the goods of development. Fossil fuels, which come from beneath the earth, have made it possible for us to forego the far more destructive, inefficient, and polluting use of wood and other so-called natural fuels that must be harvested from the earth's surface. Paradoxically, fossil fuels may have even helped save whales from extinction. Prior to learning how to use petroleum, humans had few alternatives to whale oil for generating heat and light.

Moreover, fossil fuels, such as coal and oil, have also had far-reaching positive environmental effects that a good steward should wish to consider in drawing up the global balance sheet. The first effect is to make it possible for farmers to replace beasts of burden with machines and, therefore, to cultivate land more efficiently. (Much of the developing world is now beginning to undergo this process of agricultural modernization today.) Second, fossil fuels have been turned into fertilizers that, together with new pesticides, other means of preventing spoilage, and advances in new plant species—the so-called Green Revolution—have produced so much more food per acre that large amounts of land have now been spared from cultivation altogether. For example, America's forests, contrary to popular perception, have been growing steadily for the past fifty years and are actually larger than they were one hundred years ago.[21] Even in the heavily populated coastal areas, small farms have returned to forestland. The result of all this is that despite its vast fossil-fuel consumption, North America currently shows a net minus in the amount of carbon dioxide it puts into the atmosphere. In other words, North America absorbs more carbon dioxide through plants and forests than it emits through industry.[22] No one intentionally set out to produce these consequences, but human ingenuity, aimed at doing better with greater cost efficiency and lower amounts of raw materials, seems here to reflect a providential convergence of man and nature. Now that we are conscious of the effects of our activity on nature, we can set out to do even better.

If other countries in the world could imitate such ingenuity and efficiency, we would not see an exhaustion and despoliation of natural resources. Instead, we would see their enhancement and protection. Agricultural scientists have estimated that if the rest of the world could achieve the level of efficiency and care for the land exhibited by the average farmer in the developed world, then ten billion people—which is almost twice the current world population, and is a larger figure than is now expected when the population levels off in the middle of the century—could be fed on half the land. Put into concrete terms, this means that *an area the size of India could simply be left untouched worldwide in spite of population growth.*[23] It is a modern scandal, then, that out of a misguided concern for the earth, some philanthropic foundations and environmental groups from developed countries, and some international agencies as well, have discouraged, or even refused to support so-called "unsustainable" agricultural practices. These practices are, in fact, necessary for saving and improving the lives of the world's poor and hungry.

Such a position severely clashes with the moral imperative outlined above that human needs must be given first priority in environmental policy and practice. There is room for well-meaning people to disagree about the best set of stewardship policies; and it is rarely the place of the Catholic Church to endorse particular policy proposals. However, we should not indulge ourselves in a strongly negative, almost anti-human view of human population. Unfortunately, environmental policy is often guided by this view—a view that ultimately deplores the appearance of billions of new people on the planet, each of whom, by God's providence, is created to enjoy eternal life with him. Many environmentalists seem to believe that human beings are a kind of scar or cancer on the land, an immoral intrusion on an otherwise perfect natural order. No basis for this view can be found in revelation; indeed, quite the opposite is true. Man was placed here by God and was commanded to be fertile and multiply, to fill the earth and subdue it (Gen. 1:28). Thinking of the existence of other people as unfortunate and perhaps even as a violation of nature is a radical departure from the Judeo-Christian ethic. We are made in God's image and likeness, and that means, in part, that *every* person conceived is sacred, per se, because he or she adds to creation an incommunicable value that did not previously exist. The view that people are merely a drain on resources not only contradicts our faith, but denies the real contributions of human beings to the common good of human society and the integrity of the environment. God has decided to allow these new persons into the cosmic community of spirits. Any view that does not welcome human beings both in themselves, and for what they may providentially bring into the world, is at fundamental odds with the Catholic ethos.

In addition, the best evidence appears to suggest that no population crisis, as such, exists. Some countries with high population densities are poor because their economic development has not, in fact, matched growth in human numbers. However, countries such as Japan and Hong Kong show that such poverty is an economic rather than a population problem. We have already seen that there is no shortage of food on the planet. There is equally no "population bomb" ready to go off. The predictions of alarmists on this score in the 1960s and 1970s proved false. Only nature or the disregard for human life has produced large numbers of human deaths in recent decades. Globally, food production has outstripped population growth, thanks to human innovation.

However, many human beings still suffer from a lack of basic necessities. Thus, if there does exist an imbalance between population and the amount of arable land, observes Pope John XXIII, "necessity demands a cooperative effort on the part of the people to bring about a quicker exchange of goods, or of capital, or the migration of people themselves."[24] Thus, an approach that favors economic development and international cooperation should be promoted as an alternative to programs intended to reduce human population.

Another side effect of development—albeit an unintended one—has appeared as well. As food becomes more plentiful and medicine more widely available, population growth naturally slows. Many developed countries in North America, Europe, and Asia are actually facing precipitous population collapse, absent immigration.[25] In developing countries, population growth slows as people become confident that, thanks to material improvements, more of their children will survive into adulthood. Whereas a half century ago, women in developing countries had to bear, on average, six children to keep the population steady, today's lowered infant mortality rate has cut the number of births in half.[26] Developing countries today are at the stage of many developed countries more than fifty years ago, with the added advantage of developed technologies and practices already discovered and in use. Thus, addressing the needs of developing nations is well within our potential.

What may block the path to development, however, is mistrust of human innovation, and the inevitable drags on progress that government management of the economy, weak protection of private property rights, and barriers to trade introduce. We know from hard historical experience, for instance, that the centralized, planned systems of the former

Communist countries were poor stewards of lands with remarkable natural resources. These countries were not only terribly inept at producing and distributing goods to their own people, they were also among the worst polluters and most reckless environmental regimes in history.[27] Despite many laws stipulating production targets and pollution controls, scarcity and environmental degradation were the result. Command economies and the rigidity they introduce into social relations make the environment a marginal concern. Most government planning tends to produce exactly the opposite of what is intended by hampering or penalizing needed innovations and the dynamic spontaneity to solve problems in both the economic and environmental spheres.

It is a normative Catholic principle that God intended the goods of the earth for the benefit of all.[28] In other words, while private property, as Saint Thomas Aquinas notes, is a right, it is not an absolute right.[29] Unfortunately, recent attempts to promote the common good by overly centralized planning remind us that, other things being equal, the right to economic initiative and the natural interest we take in our own property play an important social function in both the economy and the environment.

VII. A Proper Understanding of Environmental Stewardship

What becomes clear to us in this analysis is that we need a very sophisticated grasp of our situation that will take into account everything that the sciences—which are a product of human reason—are able to tell us about our world. Yet this is not all; we must also integrate our scientific knowledge with the normative principles of the moral order.

The moral teaching of the Church, as manifested in the various saintly lives of Christians throughout history, remains a key component in our understanding of how we should live in relationship to the material world. These individuals have challenged us to see that it is prudent for us—as both bodies and spirits—to refrain from consuming more than we need, or to coarsen ourselves by the endless pursuit of luxuries. Our tradition challenges us to be very careful in our personal lives about the temptations of worldly goods. Yet what is helpful, and even a religious necessity, in one's personal life cannot be translated directly into a social ethic without some caveats.

The human species as a whole will do better for itself and for creation if we vigorously cultivate the intelligence and creativity with which we have been endowed. This can be accomplished when each person is allowed the economic freedom to seek material improvements, and to make them economically viable within a system that is circumscribed by a strong juridical framework.[30] A more expansive social ethic that allows for economic prosperity does not contradict personal austerity, as it may appear at first glance. Large-scale innovation and productivity actually allow for greater efficiency, thus saving raw materials and energy in the long run. As the Catholic tradition acknowledges, proper distinctions are an imperative for moral analysis. Thus, it may be important to generate a lot of wealth; however, what one does with that wealth is quite another matter.

Moreover, while we ought to desire a certain simplicity in our personal lives, returning to some pre-industrial agrarian arrangement would result in the loss of such goods as profitable employment, modern medicine, and a resilient infrastructure, as well as in reduced food production, thus creating an empty well of human need. In times past, human existence was marked by a constant struggle for survival. Only since industrialization has man acquired the means necessary for protecting himself against the forces of nature. Putting the billions of people now alive back on the land would, paradoxically, have even worse environmental effects than intelligent development. Consequently, economic development must progress hand in hand with individual commitment to the virtue of temperance.

Similarly, no responsible person believes that the relatively simpler but dirty old path of early industrialization should be continued in the future. Many environmental problems are already well on their way to technical solutions. Water and air are vastly cleaner than they were only two decades ago, largely due to advances in technology. Manufacturing processes and automobiles may soon have no environmental effects whatsoever. Thus, in addition to the great advances we have seen in agriculture and medicine, we can anticipate that, in the very near future, technologies will continue to provide ways to solve many other problems we currently face.[31] However, to achieve a reduction in environmental impact, human societies require greater development and more innovation, not less.

Since questions of stewardship, by their nature, reflect great human as well as natural complexity, public policy must reflect the greatest technical skill, practical wisdom, and widest human experience possible. Experience has shown that democratic political systems and market economies, by and large, do exactly that, particularly when moral values and the practice of virtue inform them. Neither of these systems is perfect, and neither will deal with the environment perfectly. Both are subject to the pitfalls of human vice, fallibility, and original sin—as well as simple error. However, as Thomas Jefferson observed, there is "no safe depository of the ultimate powers of the society but the people themselves." Time has proven the practical wisdom of that principle, and we might observe that it is consistent with the Catholic view that every human person has been endowed by God with gifts intended to be used for the glory of the Creator and of his creation. Democracy and a free economy provide a space for those gifts to be effectively utilized in the stewardship of the earth. It is often argued that environmental questions are so urgent that they cannot wait for a popular consensus to form or cannot depend upon market incentives—which are often focused on short-term gains—for solutions. In a very few cases of demonstrated emergency, that may be true. In almost every other instance, however, far from being inconvenient obstacles to realizing environmental goals, democracy and markets are the most effective social embodiments of our God-given intelligence, and are the best mechanisms for the responsible handling of the environment.

It is no mere coincidence that the words *ecology* and *economics* have related etymologies. *Economics*, referring to the laws of the household (*oikos* in Greek), is the science of how we produce, sell, buy, trade, and use goods and employ services to meet human needs. *Ecology*, a word that came into existence in the nineteenth century as environmental questions became more evident, is the science of the laws that govern the interactions of the earth's biosphere with the earth's inhabitants, specifically as the home (*oikos*) for all life (*bios*). The two terms are deeply related in reality as well as in their origins. Too often, however, they are set in opposition to each other. The usual way this relationship is characterized is by arguing that greed, expressed in economic activity, is the driving force behind ecological problems. Even historically, this is false. The economic actions intended to fulfill human needs have often damaged ecological systems, but to portray these actions simply as greed or excessive consumption assumes that nature is far more benign than the witness of human history seems to suggest. Much of the environmental harm inflicted on nature in the past few centuries has stemmed from human ignorance, not malice or even greed, as we have tried to gain advantage over the nemesis of material scarcity. Yet now that we are beginning to discern the value of our stewardship over nature, we are in a new situation. Thus, we need to reaffirm our commitment to the tools that allow us to respond effectively to the multifaceted problems we face.

First, we need *the very best and dispassionate environmental science* to help us sort out the immensely complicated series of interconnected effects of our actions on the biosphere. Simple emotional appeals or alarmist claims are of little use here. As Pope John Paul II has pointed out, "Reverence for nature must be combined with scientific learning."[32] Global warming, for instance, which remains speculative and based on incomplete computer models rather than on demonstrated science, might cost man and nature a great deal if we rush to impose dramatic limits on fossil-fuel use in a misguided attempt to solve a problem that may not even exist. Just twenty-five years ago, some of the current proponents of global warming were warning us about global cooling.[33] Because ecology is still in its infancy, we need to utilize all that we know to help us find prudent solutions for these complex problems. We must also recognize that science alone is insufficient for resolving these matters, especially since these issues have moral implications. Thus, in recognizing that we will have to make unavoidable tradeoffs in striking a balance between human need and a clean environment, we must exercise prudence in addressing environmental concerns.

Finding ways for nature and man to coexist for the benefit of all of creation will demand great human ingenuity and effort in the coming years. At the moment, the simplest solution for many environmental problems is to set aside land for conservation and wildlife habitat. Around the world, the countries that enjoy the greatest prosperity are able, through both public and private means, to set aside land for wildlife preservation. Development and wealth make environmental care much easier, as can be inferred from the fact that intelligent development, which leads to a surplus of wealth, provides the greatest possibility for man to address concerns beyond the scope of his immediate material needs. This fact is rooted in the very logic of man's dominion over nature.[34] Despite some continuing environmental problems, developed countries are the ones most dedicated to and successful in treating their own environmental situations.

For the most part, it is not the entrepreneur or the corporation in developed societies, as is often claimed, who acquires disastrous and short-term profit at the expense of the environment. Entrepreneurs usually have a vested interest in their own kind of sustainability, as well as incentives to innovate and to make products more efficiently and with less waste. By contrast, the poorest and least-developed countries frequently have few real options as their often-growing populations, with little or no incentive to prudent stewardship of their natural resources, exploit every available resource in the search for short-term survival.

The poorer countries of the world are those most in need of good science and development, for both economic and environmental reasons. The traditional forms of agriculture and manufacturing, often romantically thought to be ideal models of how to live on the land, are actually a much heavier burden upon earth and upon man than modern developments. For example, developing countries would benefit both environmentally and economically from electricity. Electricity generated by fossil fuels is frequently portrayed as a clumsy and centralized means of power generation that would best be replaced by wind, solar, or wave-powered generators. If these alternative energy sources were successfully developed and made affordable, perhaps this would be true. However, in the meantime, millions of children and adults die every year in developing countries because of the smoke they inhale from wood and dung fires, or because of the impure water that they must drink for lack of proper sanitation. Thus, their basic needs would be met with far less local and atmospheric pollution by the construction of the most up-to-date electrical power generators around the world. Even if this source of energy is not perfect, it represents an improvement toward both meeting human need and a cleaner environment. Science and development should work in tandem to aid the most hard-pressed of our human neighbors, while taking prudential steps toward a fuller realization of environmental stewardship.

In addition to proper science, however, we desperately need an *authentic democratic deliberation* on the environment. Every recent survey of the American people confirms that they place high value on a clean and safe environment. Yet in human life there are few indisputable absolutes. Thus, we see that most often these same people do not endorse the proposals recommended by many environmental organizations for achieving this seemingly desired end. Real environmental decisions, as we have seen, always involve choices between different and sometimes competing values, therefore suggesting that we must proceed with great caution and prudence.

For example, air quality in the United States is better than it has been in decades. Soon, smog is likely to be a thing of the past. Pollutants are still put into the atmosphere by human activity, but, at a certain point, a moral calculation must be made. Do we want to spend enormous amounts of human and material capital on removing, say, the last 5 percent of an air pollutant at the cost of being unable to deal with other more serious problems? If so, what if the last 2 percent is ten times as expensive? Or a hundred times? Prudence dictates that we need a moral and political calculus that will weigh several competing values as they come to bear upon the common good. Though all of them are perhaps laudable enough in themselves, we must always consider the fact of scarcity when seeking to resolve these conflicts of interest. By virtue of the limits placed on our material existence, we must be modest in our assessment of what we can reasonably achieve environmentally without placing an undue hardship on others. True democratic processes, then, will allow for the real cost and benefits of environmental stewardship to emerge, and thus a policy can be advanced that truly upholds the common good.

Third, in much of the literature on the environment, entrepreneurs and the technologies they employ are pitted against ecologists and the "rights" of nature. There is a kernel of truth in such arguments, because all human activity alters the natural world to a greater or lesser degree. Far from being locked in inevitable conflict, however, *entrepreneurs and environmentalists need increasingly to cooperate* with one another to the benefit of both. Many environmentalists have demonized entrepreneurs. Without going to the opposite extreme of idealizing entrepreneurs—some of whom provide great service, others of whom, in fact, are irresponsible—it is clear that there are several ways in which entrepreneurial activity, at its best, will be crucial to the solution of environmental problems. First, scientific research, both in nonprofit and in corporate settings, depends largely on the excess capital generated by successful entrepreneurs. Entrepreneurs also have a market incentive in developing innovations favorable to the environment, such as new technologies that replace older, dirtier, and less efficient technologies. Only the freedom and responsiveness of markets, as has been demonstrated around the world, will succeed in distributing those goods to the widest number of people. As Pope John Paul II has argued, "the free market is the most efficient instrument for utilizing resources and effectively responding to needs."[35] Environmentalists can play a useful role in identifying problems and threats. However, as it stands today, their critiques are often insufficient for addressing the vast array of needs confronting society as a whole. Therefore, embracing a broader view of creation that credits economic activity as being an extension of God's own wisdom for how man is to relate to his physical surroundings is becoming increasingly important.

Fourth, many environmentalists deplore the right to private property. In contrast, property is upheld in the Catholic tradition, not only as a fundamental right by virtue of man's labor, but also as the means by which God intends man to develop the earth for the benefit of all people. Property that is held in common is most often neglected. In general, he who owns his property will care for it and produce something from it. Therefore, an owner is typically the best steward of a resource. However, the right to private property, in Catholic social thought, can never be understood authentically apart from the universal destination of all material goods. Man is entitled to the fruits of his labor, only inasmuch as he has a right to provide for himself and his family, and the duty to help others in need. Saint Thomas Aquinas provides several arguments for why privately owned property is better cared for than common property or property that is owned by no one in particular.[36] In short, he argues that property is temporary and relative in this world. Since its possession requires moral as well as legal limitations, where private property rights have been respected, the whole created order has generally fared better.

Some environmental problems may be *best* treated, in fact, by creating new forms of property rights defensible by law. The law has recognized that pollution damages the common environment and may, therefore, be curtailed in respect to others' property rights. Recently, pollution credits, which are currently being actively traded, have provided successful market incentives to reduce emissions. However, we have not yet experimented extensively with ways to use private-property rights to solve ecological questions. Nonetheless, limited experimentation in this area has yielded positive results. For some places in Africa, for example, establishing property rights over land and animals, and allowing local peoples to benefit from controlled hunting and harvesting policies, have paradoxically lessened poaching and made hunting both economically valuable and sustainable. Previously, people in such areas had immediate incentives to destroy large beasts and their habitats in order to enlarge simple agricultural activities. Innovation that takes advantage of new markets has enabled them to avoid harming nature, to a greater degree, while also benefiting themselves. Whenever possible, as this example illustrates, economic and ecological interests must everywhere be made to coincide as closely as possible with one another.

VIII. Recommendations

In conclusion, we would like to recommend some general principles as guides to future reflection on environmental questions:

1. *Nature reveals God as the Creator.* Thus, we human beings learn things about God and ourselves from contemplating the earth's power, intelligibility, and beauty. We would do well to know nature better in its immediacy and to cultivate the ancient practice of meditating on nature in order to increase our spiritual understanding and love for God's world. As Pope John Paul II rightly reminds us, "Our very contact with nature has a deep restorative power."[37]

2. *Even natural contemplation, however, will lead us, as it did many early civilizations, to see that nature points to something beyond itself and draws man to the ultimate source of well-being.* We care for creation as a God-given responsibility, but the love of neighbor as a being with an eternal destiny is a still higher demand. We should welcome new additions to our numbers by protecting the sanctity of human life—from conception to natural death—and taking all possible steps to see that each person's basic needs are met. The United States Catholic Conference has posed this question: If we do not respect human life, "can we truly expect that nature will receive respectful treatment at our hands?"[38]

3. *Meeting human needs should not be thought of as a zero-sum process that inevitably entails further deterioration of nature or exploitation of neighbor.* Creative minds and ready hands can quite easily offset and even reduce the current human impact on creation and can expand man's capacity to meet the needs of his neighbor through voluntary exchange.

4. *Ecology and economics must go hand in hand.* (Sound environmental stewardship is the joining of the two.) There is an economy of salvation, an economy of human existence, and an economy of the environment. Greater prosperity generally correlates with greater concern for—and better means for dealing with—environmental questions. It also leads to voluntary, non-coercive decisions about having children—decisions that avoid morally illicit means of reducing perceived population pressures.

5. *Political and economic liberty best reflect the human freedom and intelligence with which we have been endowed by God.* Democratic political systems and free economies, therefore, are most likely to respond to our environmental concerns in the most fully human way. In many cases, this

means that finding market solutions to perceived problems will benefit both people and the environment.

6. *We should resist the tendency to believe that centralized planning is more environmentally responsible than free institutions.* The countries that have had the most centralized systems in the past century have also been the most harmful to the environment. Catholics are not opposed to properly constituted state power, but the issues where clumsy and rigid regulation can help are far fewer than is generally believed. Agile and flexible markets can respond, and with great efficiency, to problems unsolvable in any other way.

7. *Entrepreneurship is one dimension of human nature.* Portraying all entrepreneurs as people driven merely by greed is both unfair and disrespectful to one of the means God has given us to handle our ever-changing needs. Properly understood, responsible entrepreneurship is a vehicle for realizing what the United States Catholic Conference has called a "common and workable environmental ethic."[39] As Pope John II has stated, "Placing human well-being at the center of concern for the environment is actually the surest way of safeguarding creation; this in fact stimulates the responsibility of the individual with regard to natural resources and their judicious use."[40]

Conclusion

The revelation of God both in nature and in salvation history does not lead us to believe that we should return to some prelapsarian garden in the earth's distant past. Angels with flaming swords block that way forever (Gen. 3:24). As Pope John Paul II has pointed out, ecological responsibility "cannot base itself on the rejection of the modern world or on the vague wish for a return to a 'lost paradise.'"[41] Human dominion over nature is not necessarily evil; yet our task lies before us. We must always be on guard against a two-fold temptation that is repeatedly denounced by God: first, making idols of nature or creatures that, in so doing, exalts them above our primary duties toward God; and, second, neglecting the needs of our human neighbor. We are awaiting the New Jerusalem, a city to be given to us at the end of time out of God's free bounty, which will descend upon a New Heaven and a New Earth. In the meantime, we must combat the evil in ourselves and in our world. We must seek better ways to love God by keeping his commandments and loving our neighbor as ourselves. In a sense, the love for our neighbor can be extended to the non-human world. However, we will have to make prudential judgments about many complex questions and expect inescapable tradeoffs along the way. Since "one can love animals" but should not "direct to them the affection due only to persons,"[42] whenever there is an unavoidable choice between people and nature, we must, like God, put people, the summit of his creation, first.

Finally, we should always have faith that God never abandons his people. Our talents were given to us for a reason: to enable us to love God and our neighbor in Christian freedom. We may be confident that God will also provide us with the gifts and graces that are needed to care for both nature and ourselves. Nonetheless, we should still not expect that any of our many pursuits in the coming years—let alone complex activities such as environmental stewardship—will be without new problems of their own. As the great Catholic theologian Hans Urs von Balthasar has recently reminded us, Jesus said that the wheat and the tares grow together. Believing that we can uproot all evil may threaten the goods on which we all depend.[43] Catholic teaching about the Fall is a realistic, not a pessimistic view, in this perspective. There is much bad and much good in our world, but the persistence of evil should not discourage us. Until the Lord comes in glory, total perfection for us as a species and perfect harmony within nature are beyond our reach, but we know that someday he will come. In the meantime, we seek salvation and our human future amid great uncertainties, but also in joyful hope that the Creator who brought this world and the human race into being is certainly still at work in it—and in us.

Editorial Board

Notes

1. *Catechism of the Catholic Church* (1994), 385.

2. Ibid., 343.

3. Cf. Aristotle, *Metaphysics* 1.2.

4. Saint Bonaventure, *Legenda Major* 4.3. See also Omar Englebert, *Saint Francis of Assisi: A Biography* (Chicago: Franciscan Herald Press, 1965).

5. Cf. Frederick Copleston, S.J., *A History of Philosophy*, vol. 2, part 1 (New York: Image, 1963), 164.

6. The Second Council of the Vatican, *Lumen Gentium* (November 21, 1964), 36.2.

7. *Catechism of the Catholic Church* (1994), 397–398.

8. Ibid., 400.

9. United States Catholic Conference, Pastoral Statement Renewing the Earth: An Invitation to Reflection and Action on Environment in Light of Catholic Social Teaching (November 14, 1991), III, A.

10. Cf. *Catechism of the Catholic Church* (1994), 342.

11. United States Catholic Conference, Pastoral Statement *Renewing the Earth: An Invitation to Reflection and Action on Environment in Light of Catholic Social Teaching* (November 14, 1991), II, A.

12. Saint Augustine, *City of God* 12.20.

13. Saint Augustine, *Epistles* 138.1.

14. Cf. Alan Macfarlane, *The Culture of Capitalism* (Oxford: Basil Blackwell, 1987).

15. John Paul II, Encyclical Letter *Centesimus Annus* (May 1, 1991), 31.

16. *Catechism of the Catholic Church* (1994), 339–340.

17. Paul VI, Encyclical Letter *Octogesima Adveniens* (May 14, 1971), 21.

18. John Paul II, "The Ecological Crisis: A Common Responsibility," *1990 World Day of Peace Message* (December 8, 1989), 13.

19. *Catechism of the Catholic Church* (1994), 339.

20. Ibid., 2415.

21. See Jesse H. Ausubel, "The Liberation of the Environment," *Daedalus* 125 (summer 1996): 1–17.

22. S. Fan, M. Gloor, J. Mahlman, S. Pacala, J. Sarmiento, T. Takahashi, and P. Tans, "A Large Terrestrial Carbon Sink in North America Implied by Atmospheric and Oceanic Carbon Dioxide Data and Models," *Science* 282 (October 16, 1998): 442–446.

23. Paul E. Waggoner, "How Much Land Can Be Spared for Nature?" *Daedalus* 125 (summer 1996): 87.

24. John XXIII, Encyclical Letter *Pacem in Terris* (April 11, 1963), 101.

25. Cf. Nicholas Eberstadt, "World Depopulation: Last One Out Turn Off the Lights," *Millken Institute Review* 2 (first quarter 2000): 37–48.

26. United Nations, *World Population Prospects: The 1994 Revision.*

27. Murray Feshbach and Alfred Friendly, Jr., *Ecocide in the USSR: Health and Nature Under Siege* (New York: Basic Books, 1992).

28. *Catechism of the Catholic Church* (1994), 2401.

29. Saint Thomas Aquinas, *Summa Theologiae* II–II Q. 66.

30. John Paul II, Encyclical Letter *Centesimus Annus* (May 1, 1991), 42.

31. Paul Hawken, Amory Lovins, L. Hunter Lovins, *Natural Capitalism: Creating the Next Industrial Revolution* (Boston: Little, Brown, and Company, 1999).

32. United States Catholic Conference, Pastoral Statement Renewing the Earth: An Invitation to Reflection and Action on Environment in Light of Catholic Social Teaching (November 14, 1991), IV, B.

33. Anna Bray, "The Ice Age Cometh: Remembering the Scare of Global Cooling," *Policy Review* 58 (fall 1991): 82–84.

34. Cf. Gene M. Grossman and Alan B. Krueger, "Economic Growth and the Environment," *Quarterly Journal of Economics* 110 (May 1995): 353–377.

35. John Paul II, Encyclical Letter *Centesimus Annus* (May 1, 1991), 34.

36. Cf. Saint Thomas Aquinas, *Summa Theologiae* Ia–IIae, q. 105, aa. 2–3, and IIa–IIae q. 66.

37. John Paul II, "The Ecological Crisis: A Common Responsibility," *1990 World Day of Peace Message* (December 8, 1989), 14.

38. United States Catholic Conference, Pastoral Statement Renewing the Earth: An Invitation to Reflection and Action on Environment in Light of Catholic Social Teaching (November 14, 1991), III, H.

39. Ibid., I, D.

40. John Paul II, "Respect for Human Rights: The Secret of True Peace," *1999 World Day of Peace Message* (January 1, 1999), 10.

41. John Paul II, "The Ecological Crisis: A Common Responsibility," *1990 World Day of Peace Message* (December 8, 1989), 13.

42. *Catechism of the Catholic Church* (1994), 2418.

43. Hans Urs von Balthasar, *A Theology of History* (New York: Sheed and Ward, 1963), 124–125.

Part 4

A Biblical Perspective on Environmental Stewardship

Introduction

In the last three centuries, life expectancy in advanced economies has risen from about thirty years to nearly eighty. Cures have been found to once-fatal diseases, and some diseases have been eliminated entirely. Famine, which once occurred, on average, seven times per century in Western Europe and lasted a cumulative ten years per century, is now unheard of there. While the average Western European family in A.D. 1700 lived in a hovel with little or no furniture, no change of clothing, and barely enough food to sustain a few hours' agricultural labor per day[1]—and, of course, they also lacked electricity, plumbing, water and sewage treatment, and all the appliances we often take for granted—today the average family lives in a well-built home with all those amenities, along with enough food to make obesity, not hunger, the most common nutritional problem even among the "poor."[2] Such advances in the West have been the fruits of freedom, knowledge, and hard work—all resting substantially on the foundation of biblical Christianity's worldview and ethic of service to God and neighbor.[3] These advances have also given rise to a laudable expansion in people's focus on the need for environmental stewardship. For as people come to feel more secure about their basic needs, they begin to allocate more of their scarce time, energy, and resources to attaining formerly less urgent ends. Consequently, the movement for environmental protection has grown as Western wealth has grown, giving rise to a strong environmental consciousness and to protective environmental legislation.

The world's less developed countries, where material progress began much later, have been catching up in the past century, as shown especially by rapidly rising life expectancy (from about thirty years in 1900 to about sixty-three years today).[4] Nonetheless, in many developing countries, the basics of sufficient and pure water and food, along with clothing, shelter, transportation, health care, communication, and so forth, still remain elusive for many people. For them, continued economic advance is crucial for health and even for life itself: It is small wonder that their attention focuses more on immediate consumption needs than on environmental protection. Tragically, however, people with a strong environmental consciousness who live predominantly in Western countries sometimes seek to impose their own environmental sensibilities on people still struggling to survive. In fact, further advances in human welfare for the poor are now often threatened by a belief in the West that human enterprise and development are fundamentally incompatible with environmental protection, which is seen by some as the quintessential value in evaluating progress. This false choice not only threatens to prolong widespread poverty, disease, and early death in the developing world, but also undermines the very conditions essential to achieving genuine environmental stewardship.

In this essay, we shall present theological and ethical foundations we believe are essential to sound environmental stewardship; briefly review the human progress erected on those foundations; and discuss some of the more important environmental concerns—some quite serious, others less so—that require attention from this Christian perspective. We shall also set forth a vision for environmental stewardship that is wiser and more biblical than that of mainstream environmentalism, one that puts faith and reason to work simultaneously for people and ecology, that attends to the demands of human well-being and the integrity of creation.

Such an approach to environmental stewardship will, we believe, promote human justice and shalom, as well as the well-being of the rest of God's creation, which his image-bearers have been entrusted to steward for his glory.

I. Theological and Ethical Foundations of Stewardship

God, the Creator of all things, rules over all and deserves our worship and adoration (Ps. 103:19–22). The earth, and, with it, all the cosmos, reveals its Creator's wisdom and goodness (Ps. 19:1–6) and is sustained and governed by his power and lovingkindness (Ps. 102:25–27; Ps. 104; Col. 1:17; Heb. 1:3, 10–12). Men and women were created in the image of God, given a privileged place among creatures, and commanded to exercise stewardship over the earth (Gen. 1:26–28; Ps. 8:5). Fundamental to a properly Christian environmental ethic, then, are the Creator/creature distinction and the doctrine of humankind's creation in the image of God. Some environmentalists, especially those in the "Deep Ecology" movement, divinize the earth and insist on "biological egalitarianism," the equal value and rights of all life forms, in the mistaken notion that this will raise human respect for the earth. Instead, this philosophy negates the biblical affirmation of the human person's unique role as steward and eliminates the very rationale for human care for creation. The quest for the humane treatment of beasts by lowering people to the level of animals leads only to the beastly treatment of humans.[5]

The image of God consists of knowledge and righteousness, and expresses itself in creative human stewardship and dominion over the earth (Gen. 1:26–28; 2:8–20; 9:6; Eph. 4:24; Col. 3:10). Our stewardship under God implies that we are morally accountable to him for treating creation in a manner that best serves the objectives of the kingdom of God; but both moral accountability and dominion over the earth depend on the freedom to choose. The exercise of these virtues and this calling, therefore, require that we act in an arena of considerable freedom—not unrestricted license, but freedom exercised within the boundaries of God's moral law revealed in Scripture and in the human conscience (Exod. 20:1–17; Deut. 5:6–21; Rom. 2:14–15). These facts are not vitiated by the fact that humankind fell into sin (Gen. 3). Rather, our sinfulness has brought God's responses, first in judgment, subjecting humankind to death and separation from God (Gen. 2:17; 3:22–24; Rom. 5:12–14; 6:23) and subjecting creation to the curse of futility and corruption (Gen. 3:17–19; Rom. 8:20–21); and then in restoration, through Christ's atoning, redeeming death for his people, reconciling them to God (Rom. 5:10–11, 15–21; 2 Cor. 5:17–21; Eph.

2:14–17; Col. 1:19–22), and through his wider work of delivering the earthly creation from its bondage to corruption (Rom. 8:19–23). Indeed, Christ even involves fallen humans in this work of restoring creation (Rom. 8:21). As Francis Bacon put it in *Novum Organum Scientiarum* (*New Method of Science*), "Man by the Fall fell at the same time from his state of innocence and from his dominion over creation. Both of these losses, however, can even in this life be in some parts repaired; the former by religion and faith, the latter by the arts and sciences."[6] Sin, then, makes it difficult for humans to exercise godly stewardship, but the work of Christ in, on, and through his people and the creation makes it possible nonetheless.

When he created the world, God set aside a unique place, the Garden of Eden, and placed in it the first man, Adam (Gen. 2:8–15). God instructed Adam to cultivate and guard the Garden (Gen. 2:15)—to enhance its already great fruitfulness and to protect it against the encroachment of the surrounding wilderness that made up the rest of the earth. Having also created the first woman and having joined her to Adam (Gen. 2:18–25), God commanded them and their descendants to multiply, to spread out beyond the boundaries of the Garden of Eden, and to fill, subdue, and rule the whole earth and everything in it (Gen. 1:26, 28). Both by endowing them with his image and by placing them in authority over the earth, God gave men and women superiority and priority over all other earthly creatures. This implies that proper environmental stewardship, while it seeks to harmonize the fulfillment of the needs of all creatures, nonetheless puts human needs above non-human needs when the two are in conflict.

Some environmentalists reject this vision as "anthropocentric" or "speciesist," and instead promote a "biocentric" alternative. But the alternative, however attractively humble it might sound, is really untenable. People, alone among creatures on earth, have both the rationality and the moral capacity to exercise stewardship, to be accountable for their choices, to take responsibility for caring not only for themselves but also for other creatures. To reject human stewardship is to embrace, by default, no stewardship. The only proper alternative to selfish anthropocentrism is not biocentrism but theocentrism: a vision of earth care with God and his perfect moral law at the center and human beings acting as his accountable stewards.[7]

Two groups of interrelated conditions are necessary for responsible stewardship. In one group are conditions related to the freedom that allows people to use and exchange the fruits of their labor for mutual benefit (Matt. 20:13–15). These conditions—knowledge, righteousness, and dominion—provide an arena for the working out of the image of God in the human person. In another group are conditions related to responsibility, especially to the existence of a legal framework that holds people accountable for harm they may cause to others (Rom. 13:1–7; Exod. 21:28–36; 22:5–6). These two sets of conditions provide the safeguards necessitated by human sinfulness. Both sets are essential to responsible stewardship; neither may be permitted to crowd out the other, and each must be understood in light of both the image of God and the sinfulness of man.

Freedom, the expression of the image of God, may be abused by sin and, therefore, needs restrictions (1 Pet. 2:16); but governmental power, necessary to subdue sin and reduce its harm, must be exercised by sinful humans, who may also abuse it (Ps. 94:20; 1 Sam. 8). This means that it, too, needs restrictions (Acts 4:19–20; 5:29). Such restrictions are reflected not only in specific limits on governmental powers (Deut. 17:14–20), but also in the division of powers into judicial, legislative, and executive (reflecting God as Judge, Lawgiver, and King [Isa. 33:22]); the separation of powers into local and central (exemplified in the distinct rulers in the tribes of Israel and the prophets or kings over all Israel [Deut. 1:15–16]); the gradation of powers from lesser to greater (Exod. 18; Deut. 16:8–11); and the vesting of power in a people to elect their rulers (Deut. 1:9–15; 17:15). All of these principles are reflected in the Constitution of the United States. Also crucial to the Christian understanding of government is the fact that God has ordained government to do justice by punishing those who do wrong and praising those who do right (Rom. 13:1–4; 1 Pet. 2:13–14).[8]

These principles indicate that a biblically sound environmental stewardship is fully compatible with private-property rights and a free economy, as long as people are held accountable for their actions. Stewardship can best be accomplished, we believe, by a carefully limited government (in which collective action takes place at the most local level possible so as to minimize the breadth of harm done in case of government failure) and through a rigorous commitment to virtuous human action in the marketplace and in government.

These principles, when applied, promote both economic growth and environmental quality. On the one hand, there is a direct and positive correlation between the degree of political and economic freedom and both the level of economic attainment and the rapidity of economic growth in countries around the world. The 20 percent of the world's countries with the greatest economic freedom produce, on average, over ten times as much wealth per capita as the 20 percent with the least economic freedom, and while the freest countries enjoyed an average 2.27 percent annual rate of growth in real gross national product per capita through the 1990s, the least-free countries experienced a decline of 1.32 percent per year.[9] On the other hand, there is also a direct and positive correlation between economic advance and environmental quality.[10] The freer, wealthier countries have experienced consistent reductions in pollution and improvements in their environments, while the less free, poorer countries have experienced either increasing environmental degradation or much slower environmental improvement. We shall return to this correlation shortly; first, however, it behooves us to know something of the changes in our material condition over the last few centuries.

II. The Marvels of Human Achievement

Until about 250 years ago, everywhere in the world, the death rate was normally so close to the birth rate that population grew at only about 0.17 percent per year,[11] doubling approximately every 425 years, instead of every forty-two years at the world's average growth rate in the 1980s, or every fifty-one years at the average rate for the 1990s.[12] Infant and child mortality rates (around 40 percent overall) were little better for the very rich—royalty and nobility—than they were for farmers and peasants, even into the eighteenth century. Britain's Queen Anne (1665–1714), for instance, was pregnant eighteen times; five of her children survived birth; none survived childhood.

Eighteenth-century French farming—the best in Europe—produced only about 345 pounds of wheat per acre; modern American farmers produce 2,150 pounds per acre, about 6.2 times as much.[13] Early-fifteenth-century French farmers produced about 2.75 to 3.7 pounds of wheat per man-hour, and the rate fell by about half over the next two centuries;[14] modern American farmers produce about 857 pounds per man-hour[15]—about 230 to 310 times as much as their French counterparts around 1400, and 460 to 620 times as much as French farmers around 1600. (This means that modern farmers also manage to farm from 37 to 100 times as many acres, thanks largely to mechanized equipment and advanced farming techniques.) As the great French historian Fernand Braudel pointed out, it became very difficult to sustain life when productivity in wheat fell below

2.2 pounds per man-hour. But for most of the 350 years from 1540 to 1890, productivity in France (which, as was fairly typical of Western Europe, suffered a serious decline in productivity at the start of that period) was well below that.[16]

Such facts help to explain why earlier generations spent a major part of each day working to earn enough income just to pay for food (excluding its preparation, packaging, transport, and serving), while we spend far less today (under 6 percent of total consumer expenditures in the United States in the 1980s went to food). These developments—along with the advent of glass window panes (to admit light and heat but exclude cold and pests) and screens (to admit fresh air and exclude disease-bearing insects); treatment of drinking water and sewage; mechanical refrigeration (to prevent food spoilage and consequent waste and disease); adoption of safer methods of work, travel, and recreation; and the advent of sanitary medical practices, to say nothing of antibiotics and modern surgical techniques—also help to explain why people live about three times as long now. While "man is destined to die once" (Heb. 9:27), the Bible recognizes death as punishment for sin and, consequently, as man's enemy (1 Cor. 15:26), and it associates long life with the blessing of God (Exod. 20:12; Deut. 11:8–9; Eph. 6:1–3) and with the reign of the Messiah (Isa. 65:20).

Economic development is a good to be sought not as an end in itself but as a means toward genuine human benefit. For instance, consider a few of the things absolutely no one—not even royalty—could enjoy before the last two centuries of economic advance:

- Electricity and all that it powers: lights, telephones, radios, televisions, refrigerators, air conditioners, fans, video cassette recorders, X-rays, MRIs, computers, the Internet, high-speed printing presses, and all other industrial automation.
- Internal combustion engines and all that they power: cars, trucks, planes, farm and construction equipment, and most trains and ships.
- Hundreds of synthetic materials such as plastic, nylon, orlon, rayon, vinyl, and the thousands of products—from grocery bags and pantyhose to compact discs and artificial body joints and organ parts—made from them.

No matter how rich people might have been a millennium—or even 150 years—ago, if they contracted a bacterial disease, they could not have been treated with antibiotics. This development was prompted by the work of the French Christian and scientist, Louis Pasteur, only in the latter half of the nineteenth century. Also, there were no more effective anesthetics than alcohol and cloves. So when limbs gone gangrenous from infections that today could be cured or, more likely, easily prevented, had to be amputated, patients gritted their teeth and hoped they would pass out from the pain of the crude saw. The germ theory of disease did not become current until the late eighteenth century, and the use of antiseptics did not begin until half a century later, with the work of the British Christian and chemist, Joseph Lister. Someone with a fever was likely to be bled to death by a doctor trying to cure it.[17]

Education was the province of the rich. Before the Reformation, few countries had widespread education, and even afterward, schooling was available principally to the rich. Two major exceptions were Germany and Scotland. In Germany, Martin Luther insisted that widespread schooling was important so that people could read the Scripture—which he had translated into the vernacular—for themselves. Similarly, in Scotland, John Knox's followers, convinced that personal knowledge of God and his Word was essential to the maintenance of civil as well as religious liberty (Ps. 119:45; Isa. 61:1; Jer. 34:15; Luke 4:18; 2 Cor. 3:17; Gal. 5:1,13; James 1:25; 1 Pet. 2:16),[18] arranged a parish-by-parish system of church-run grammar schools that ensured that practically every child could at least become literate. Scotland's high literacy rate and its Calvinist ethics of work and saving were important factors in its making contributions to the Industrial Revolution far out of proportion to its small population and earlier economic disadvantages. But even there, few were schooled for more than five or six years, and only a tiny percentage attended college, let alone graduated. Today, by contrast, in the United States, 81 percent of people twenty-five years old and over are high school graduates, and 23 percent are college graduates, and the growth in availability of education is worldwide. That is a particularly crucial factor in predicting the world's material future, because both the creation of wealth and the protection of the environment depend primarily not on brawn but on brain.[19]

The most effective measures of material welfare are mortality rates and life expectancy, because they take into account every conceivable variable that can add to or detract from a long and healthy life. A thousand years ago, human life expectancy everywhere was well under thirty years—perhaps even as low as twenty-four; today, worldwide, it is over sixty-five years, and in high-income economies, it is over seventy-six years. The under-five mortality rate has plummeted from about 40 percent everywhere as late as the nineteenth century to under 7 percent worldwide today and under 1 percent in high-income countries. And improved life expectancy comes not just from declining child mortality but from declining mortality rates at every stage of life.[20]

Materially, the world is a far, far better place today than it was not only one millennium ago, but even one century ago. Every raw material—mineral, plant, and vegetable—that plays a significant role in the human economy is more affordable (which economists recognize as meaning more abundant), in terms of labor costs, today than at any time in the past. Every manufactured product is more affordable than it has ever been.[21] And in producing all this great abundance, we have also reduced much health-threatening pollution, especially in the developed world.[22] Put simply, the world is both a wealthier and a healthier place today than ever before.

This rosy picture, however, must not generate uncritical applause for economic development, per se. Development can be positive or negative. While the fact that life expectancy keeps rising suggests that the net effect of development on human life has been positive, this does not imply that every instance of development is unmixedly beneficial, either to people or to creation. A biblical worldview and an institutional framework for prudent decision making, which we shall set forth below, are essential to ensuring that positive, rather than negative, development takes place.

We support appropriate development not for its own sake but, for example, because it uplifts the human person through work and the fruits of that labor, empowering us to serve the poor better, to uphold human dignity more, and to promote values (environmental, aesthetic, etc.) that we otherwise could not afford to promote.

The Christian tradition clearly affirms that the accumulation of material wealth should not be the central aim of life; yet people are to use wisely the gifts of creation to yield ample food, clothing, health, and other benefits. It is obvious that the great advance in wealth over the past century has taken place only in a small proportion of countries, namely, the liberal democracies and free economies of the West. Enough is now known about the administration of national economies to conclude safely that free-market systems minimize the waste of resources, and allow humans to be free and to flourish. All other systems that humans have tried lead to endless and unnecessary poverty, hunger, and oppression. For this reason, the religious communities of the Protestant tradition must take very seriously the claim that free markets and liberal democracy are essential to human welfare and therefore have a moral priority on our thinking about how society ought to be ordered.

But an ideological difficulty at present is that Western Protestant churches take too much of the present affluence for granted, misunderstand its origins, and overstate the value of the environmental amenities that have been given up to attain it. Today, this is leading many to embrace policy platforms that are explicitly against economic growth, and that give undue privilege to the preservation of the environmental status quo. This agenda threatens to deny those outside the West the very benefits that we ourselves have attained, and, ironically, it may burden the developing world with even worse environmental problems down the road. This essay challenges the arguments behind the anti-growth environmentalist agenda that is ubiquitous in today's mainstream churches, and argues that a biblical stance is entirely coherent with free-market democracy oriented toward sustainable economic growth.

III. How Economic and Environmental Trends Relate

We noted earlier that there is a direct and positive correlation between freedom and economic development and between economic development and environmental improvement. Necessarily, then, there is also a positive correlation between freedom and environmental quality. Economists find that free economies outperform planned and controlled economies not only in the production and distribution of wealth but also in environmental protection. Freer economies use fewer resources and emit less pollution while producing more goods per man-hour than less free economies. Economic demographer Mikhail Bernstam explains:

> Trends in pollution basically derive from trends in resource use and, more broadly, trends in production practices under different economic systems. In market economies, competition encourages minimization of production costs and thus reduces the use of resources per unit of output. Over time, resource use per capita and the total amounts of resource inputs also decline and this, in turn, reduces pollution....

By contrast, regulated state monopolies in socialist economies maximize the use of resources and other production costs. This is because under a regulated monopoly setting, prices are cost-based, and profits are proportional to costs. Accordingly, the higher costs justify higher prices and higher profits. This high and ever-growing use of resources per unit of output explains the high extent of environmental disruption in socialist countries.[23]

It is not only competition in free economies that encourages better stewardship of natural resources, it is also the incentive people have to protect property in which they have a financial stake. On the one hand, people naturally want their own homes and workplaces, and, by extension, their neighborhoods, to be clean and healthful, so they seek to minimize pollution. On the other hand, in a legal framework in which polluters are made liable for damage done to others' person or property, people also seek to minimize pollution that falls upon others. Moreover, a dynamic economy works to reduce pollution by finding the most efficient means of doing so. This contrasts with a command-and-control approach, in which regulators are likely to mandate particular technologies and methods for pollution control with little regard for overall social efficiency.

What we can infer from all these considerations—and what we find confirmed in empirical studies of the real world—is that free economies improve human health, raise living standards and life expectancy, and positively affect environmental conditions, doing all these things better than less free economies do. Further, the wealthier that economies become, the better they foster environmental protection. "If pollution is the brother of affluence," it has been written, then "concern about pollution is affluence's child."[24] Even if some pollution emissions rise during early economic development, the beneficial effects to human life of increased production far outweigh the harmful effects of the resulting pollution, as demonstrated in declining disease and mortality rates and in rising health and life expectancy, even during that early stage. But soon, increasing wealth enables citizens to invest more resources on environmental protection, and emission rates fall. The result has been termed the "environmental transition," which mirrors the more widely known "demographic transition."

The demographic transition is demographers' way of depicting the tendency for population growth rates to rise dramatically during early stages of economic growth and then decline back to little or no growth later. It occurs because initial increases in wealth rapidly force death rates downward in every age group, especially for infants and children, but fertility habits change only very slowly. Consequently, for a generation or two, couples continue having as many children as their forebears did, both because they expect one or two out of four children to die before maturity and also because in a primitive agricultural economy they rely upon having many young children to boost production. Then, when they become accustomed to the higher survival rates, and when the cost of raising children rises and the delay before those children become net producers rather than consumers grows, couples begin having fewer children. The result is a short-term high population growth rate preceded and followed by a long-term low (or zero) population growth rate.

Similarly, the environmental transition is a way of depicting the tendency for some pollution emissions to rise in early economic growth and then decline. Environmental economist Indur Goklany notes,

> The level of affluence at which a pollutant level peaks (or environmental transition occurs) varies. A World Bank analysis concluded that urban [particulate matter] and [sulphur dioxide] concentrations peaked at per capita incomes of $3,280 and $3,670, respectively. Fecal coliform in river water increased with affluence until income reached $1,375 per capita.
>
> Other environmental quality indicators (e.g., access to safe water and the availability of sanitation services) improve almost immediately as the level of affluence increases above subsistence. For these indicators the environmental transition is at, or close to, zero. In effect, the environmental transition has already occurred in most countries with respect to these environmental amenities because most people and governments are convinced of the public health benefits stemming from investments for safe water and sanitation. In fact, the vast majority of the three million to five million deaths each year due to poor sanitation and unsafe drinking water occur in the developing world.
>
> Other indicators apparently continue to increase, regardless of gross domestic product (GDP) per capita. Carbon dioxide and NO_x emissions and perhaps dissolved oxygen levels in rivers are in this third category. On the surface, these indicators seem not to improve at higher levels of affluence, but their behavior is quite consistent with the notion of an environmental transition. The

> transition is delayed in these cases because decision makers have only recently realized the importance of these indicators, or the social and economic consequences of controlling them are inordinately high relative to the known benefits, or both.
>
> All the evidence indicates that, ultimately, richer is cleaner, and affluence and knowledge are the best antidotes to pollution.[25]

Understanding the environmental transition, we should not be surprised to find that air, water, and solid waste pollution emissions and concentrations have been falling across the board in advanced economies around the world for the last thirty to forty years. Thus, for example, in the United States, national ambient airborne particulate emissions fell by about 80 percent from 1940 to 1994, and total suspended particulates fell by about 84 percent from 1957 to 1996; sulfur dioxide (SO_2) emissions fell by about 34 percent from 1973 to 1994, and SO_2 concentrations fell by about 80 percent from 1962 to 1996; carbon monoxide emissions fell by about 24 percent from 1970 to 1994; nitrogen oxide emissions peaked around 1972 and have declined slightly since then, while concentrations have fallen by about a third since 1974; volatile organic compounds emissions peaked in the late 1960s and by 1994 had fallen by about 30 percent; ozone concentrations fell by about 30 percent from the early 1970s to 1996; lead emissions (probably the most hazardous air pollutant) fell over 98 percent from 1970 to 1994, and concentrations also fell by about 98 percent.[26]

It is tempting to object, "This may be the case for advanced economies, but just look at the horrendous pollution in the world's poor countries!" Pollution in many of these countries is indeed horrendous. But there is no reason to think this must continue to be the case. As developing countries become wealthier—which they will do if their economic growth is not stifled by excessive government planning and by unreasonable environmental policies that suppress energy use and agricultural and industrial productivity—they have the opportunity to develop in a similar way. The environmental transition, as a concept, simply generalizes a common-sense insight: People tend to prioritize their spending in terms of their most urgent needs. Generally speaking, the most urgent material needs of the poor are for basic water, food, clothing, and shelter; in a second tier come basic health care, education, transportation, and communication; and in successive tiers come other, less urgent needs. People worried about putting food on the table today understandably consider that to be more urgent than reducing smog next year or minimizing global warming one hundred years from now. But when people are confident that their most urgent needs will be met, they begin allocating more of their resources to needs deemed by them less urgent—including increasingly rigorous environmental protection.

The rapid decline in pollution in advanced economies over the last thirty to fifty years—a decline that is continuing today—is not matched in very poor countries in early stages of economic development. But there is reason to be confident that the environmental transition not only will occur in the latter countries as surely as it has in the former, but also that it can and will occur more rapidly, with lower pollution peaks and more rapid improvements following them. Why? Because today's developing countries can cheaply import ready-made environmental protection technologies and technical know-how developed by others elsewhere at a much higher cost. That is, pollution abatement will become affordable in developing countries at much lower levels of economic development than it did in countries that progressed earlier. This is one reason trade and open dialogue between peoples are so important; they allow for the diffusion of environmentally friendly technologies and methods. The result, as illustrated in Figure 1, is a series of pollution transitions. Just as some countries went through the demographic transition long ago and others more recently, while some are in the midst of it now and others have yet to begin it, so some countries are long past the peak in the pollution transition, while others are at or just approaching it, and still others are just beginning the uptrend in pollution.

Figure 1

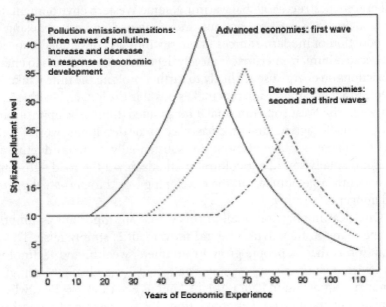

While we celebrate the decline in pollution as economies advance, however, we must not be distracted from the need to accelerate that decline in presently developing countries. Some three to five million children under the age of five die each year from diseases contracted from impure drinking water. Perhaps another three to five million die from diseases related to the widespread use of dried dung and wood for cooking and heating in the hovels of the poor, causing toxic indoor air pollution. Urban smog, largely defeated in the advanced countries of the West, remains a serious problem in many poorer cities of the world. We know how to solve these problems, as we have already done so ourselves. What the poor lack is sufficient income to afford the solutions; that is part of why economic growth in developing countries and trade between nations (which can speed the adoption of environmentally friendly technologies, management techniques, and regulatory regimes in developing countries) are so critically important—and why it is so tragic that many environmentalists embrace policies inimical to these ends. Such policies not only delay the achievement of the affluence that makes environmental protection affordable but also condemn millions of people to more years in poverty.

81

Thinking, for instance, that reducing carbon dioxide (CO_2) emissions will prevent destructive global warming, some Western environmentalists are lobbying for severe restrictions on energy use, and are opposing the introduction of modern sources of energy into less developed nations.[27] But because human enterprise is largely dependent upon access to energy, restrictions on energy use are likely to further prolong the time it takes for people to achieve the wealth that makes possible the longer, healthier lives that we in the West sometimes take for granted. Similarly, opposition to "unsustainable" agricultural practices used in the developing world—practices that serve as a take-off point for substantially more productive and environmentally sound agricultural methods down the road—threatens to condemn large numbers in the developing world to perpetual poverty and hunger.

One clear implication of all of this is that an important assumption among many in the environmental movement is simply false. The assumption is that as people grow in numbers, wealth, and technology, the environment is always negatively affected. This idea has been given formulaic expression in Paul Ehrlich's famous equation, $I = PAT$, where I is environmental damage, P is population, A is affluence, and T is technology. According to this formula, every increase in population, affluence, or technology must result in increased damage to the environment—and even more so when two or all three of these factors increase together. The damage to the environment affirmed in this vision is twofold: depletion of resources and emission of pollution. The trouble with the assumption—even though it seems intuitively sensible and certainly is a widespread belief—is that it ignores the stewardship role of the human person, and, consequently, is falsified by hard empirical data.

That pollution declines when economies grow wealthier has already been seen. The fact is illustrated well by the situation in the United States. While population grew by 19 percent from 1976 to 1994, the index of air pollution fell by 53 percent. During the same time, affluence tripled, and technology also increased dramatically, with more and more computerization and automation not only in industry and commerce but even in private homes. This is precisely the opposite of what Ehrlich's formula predicts. (See Figure 2.)

Figure 2

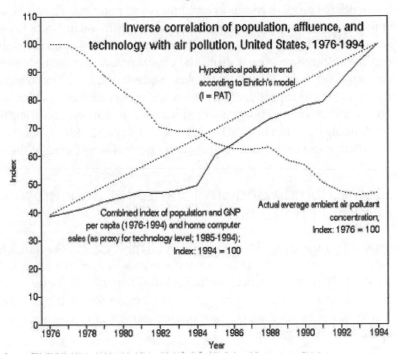

Sources: EPA; Statistical Abstract of the United States; World Bank; Social Indicators of Development on Diskette

That we are not running out of resources is also clear. Since rising prices reflect increasing scarcity and falling prices reflect decreasing scarcity, we can learn long-term resource supply trends from long-term price trends. And the long-term, inflation-adjusted price trend of every significant resource we extract from the earth—mineral, vegetable, and animal—is downward. Even more significant, the price of resources divided by wages is even more sharply downward, because while resource prices have been falling, wages have been rising. Together, these things mean that all resources are far more affordable, because they are far more abundant today than at any time in the past.[28]

Why have people so often been mistaken about the impact of growing human population and growing economies? Fundamentally, it is because they have not understood the full potential of the human person. They have considered people basically as consumers and polluters. They have not seen them—as they are presented in Scripture—as made in God's image, to be creative and productive, as he is (Gen. 1:26–28; 2:15), and as given a role in the restoration of earth from the effects of God's curse because of human sin (Rom. 8:15–25). But that biblical understanding of human nature leads Christians to expect precisely what we have seen: that, particularly when accompanied by properly formed human institutions and scientific understanding built on a biblical worldview, people can produce more than they consume and can actually improve the natural world around them.

IV. Some Human and Environmental Concerns for Present and Future

Despite the reassuring picture painted by all these general observations, many people continue to fear that we face serious threats to human well-being and to the environment as a whole. How realistic are these fears, and, to the extent that there are real dangers, what can we do about them? Let's look at three important examples: population growth, global warming, and rampant species extinction.

Population Growth

"The population crisis," writes cultural historian and evolutionary theorist Riane Eisler,

> lies at the heart of the seemingly insoluble complex of problems futurists call the world problematique. For behind soil erosion, desertification, air and water pollution, and all the other ecological, social, and political stresses of our time lies the pressure of more and more people on finite land and other resources, of increasing numbers of factories, cars, trucks, and other sources of pollution required to provide all these people with goods, and the worsening tensions that their needs and aspirations fuel.[29]

Eisler's words represent a common understanding of population growth among environmentalists: It threatens the earth with resource depletion and pollution. As we have seen, however, empirical observation, as well as biblical understanding of the implications of the image of God in the human person, suggests the opposite conclusion.

Nonetheless, many people still fear population growth because they believe it leads to overpopulation. When asked what they mean by *over-population*, they usually speak of crowding and poverty. Yet the assumption that high population density begets those things is mistaken. Some of the most desirable places to live in the world are also among the most densely populated. Manhattan, for instance, with its density of over 55,000 people per square mile, also has very high rents—a sure sign that plenty of people really want to live there, despite its high density. Or maybe, instead, they want to live there precisely because of its high density. The teeming population of Manhattan brings together a magnificent mix of human talent that makes life there fascinating, challenging, and rewarding for its millions. Similar things are true of all the world's great cities. With all their problems, they clearly attract more people than they drive away. Why should we question people's judgments about where they choose to live?

Some people think high population density lies at the root of poverty in developing nations such as China and those in sub-Saharan Africa. Yet China's population density is less than one-fifth of Taiwan's, and, aside from their forms of government, the two countries have very similar cultures. Taiwan, however, produces about five times as much wealth per capita as China. And the Netherlands, with population density nearly four times China's, produces more than ten times as much wealth per capita. And sub-Saharan Africa? Despite the common belief that it is overpopulated, it actually suffers instead from such low population density (just over half that of the world as a whole and lower than the average densities of the high-, middle-, and low-income economies of the world) that it cannot afford to build the infrastructure needed to support a strong economy.[30]

In reality, overpopulation is an empty word. As demographer Nicholas Eberstadt puts it, "the concept cannot be described consistently and unambiguously by demographic indicators." Eberstadt asks,

> What are the criteria by which to judge a country "overpopulated"? Population density is one possibility that comes to mind. By this measure, Bangladesh would be one of the contemporary world's most "overpopulated" countries—but it would not be as "overpopulated" as Bermuda. By the same token, the United States would be more "overpopulated" than the continent of Africa, West Germany would be every bit as "overpopulated" as India, Italy would be more "overpopulated" than Pakistan, and virtually the most "overpopulated" spot on the globe would be the kingdom of Monaco.

Rates of population growth offer scarcely more reliable guidance for the concept of "overpopulation." In the contemporary world, Africa's rates of increase are the very highest, yet rates of population growth were even higher in North America in the second half of the eighteenth century. Would anyone seriously suggest that frontier America suffered from "overpopulation"?

What holds for density and rates of growth obtains for other demographic variables as well: birthrates, "dependency ratios" (the proportion of children and elderly in relation to working age groups), and the like. If "overpopulation" is a demographic problem, why can't it be described unambiguously in terms of population characteristics? The reason is that "overpopulation" is a problem that has been misidentified and misdefined.

The images evoked by the term overpopulation—hungry families; squalid, overcrowded living conditions; early death—are real enough in the modern world, but these are properly described as problems of poverty.[31]

Despite all this, some people still fear population growth. Their fears, however, lack both biblical and empirical bases. First, the Bible presents human multiplication as a blessing, not a curse (Gen. 1:28; 8:17; 9:1, 6–7; 12:2; 15:5; 17:1–6; 26:4, 24; Deut. 7:13–14, cf. 30:5; 10:22, cf. 1:10; Ps. 127:3–5; 128:1, 3; Prov. 14:28); in contrast, a decline in population was one form of curse God might bring on a rebellious people (Lev. 26:22; Deut. 28:62–63). Second, although some people continue to believe projections made thirty and forty years ago of the world population topping twenty, thirty, or even forty billion in the next century or so, demographic trends indicate that the reality will be quite otherwise. Those projections were made based on the highest population growth rate the world has ever seen—about 2.2 percent per year in the 1960s, the peak of the global demographic transition. But by the year 2000, the worldwide population growth rate had dropped to about 1.3 percent per year, and it is expected to drop even further as the demographic transition plays itself out. Eberstadt explains:

Today, almost one-half of the world's population lives in 79 countries where the total fertility rates [TRFs] are below replacement (an average of 2.1 children per woman over her lifetime).... The TRFs in countries with above-replacement rates are beginning to fall. For all Asia, TRFs have dropped by over one-half from 5.7

children per woman in the 1960s to 2.8 today. Similarly, Latin America's average TRFs fell from 5.6 in the 1960s to 2.7 today. If U.N. median-variant projections of world population turn out to be correct, world population will be 7.5 billion in 2025 and 8.9 billion in 2050.

But even that might be overstating likely future population. "If present global demographic trends continue, the U.N. low-variant projections are likely. That would mean that world population would top out at 7.5 billion in 2040 and begin to decline."[32]

There is no good reason to believe that overpopulation will become a serious problem for the world. On the contrary, the more likely problem is that an aging world population will put greater stress on younger workers to provide for older, disabled persons.[33] Such a prospect, coupled with the sanctity of human life, makes all the more tragic the support in many quarters for morally illicit means of population control. Only genuine barriers to human flourishing create the problems associated with "overpopulation"; attacking problems such as poverty head-on is a far better way of improving human welfare and upholding human dignity than simply deeming certain lives unworthy of living and so, in the name of fighting "overpopulation," embracing abortion, euthanasia, and other actions that undermine the sanctity and dignity of human life.

Global Warming

Global warming is the biggest of all environmental dangers at present, maintain many environmentalists. Ironically, the great fear thirty years ago was of global cooling, for scientists recognized then that the earth is nearing a downward turn in its millennia-long cycle of rising and falling temperatures, correlated with cycles in solar energy output. But no more. Now people fear that rising atmospheric carbon dioxide, called a "greenhouse gas" because it traps solar heat in the atmosphere rather than allowing it to radiate back into space, will cause global average temperatures to rise. The rising temperatures, they fear, will melt polar ice caps, raise sea levels, cause deserts to expand, and generate more and stronger hurricanes and other storms. Are there good reasons for these fears?

While atmospheric carbon dioxide (CO_2) is certainly on the rise, and global average temperature has almost certainly risen slightly in the last 120 years or so, it is by no means certain that the rising temperature stems from the rising CO_2. The most important contrary indicator is that the sequence is the reverse of what the theory would predict. Almost all of the approximately 0.45°C increase in global average temperature from 1880 to 1990 occurred before 1940, but about 70 percent of the increase in CO_2 occurred after 1940. If the rising CO_2 was responsible for the rising average temperature, the reverse should have been the case. In addition, roughly two-thirds of the overall increase is attributable to natural, not manmade, causes (primarily changes in solar energy output).[34]

Highly speculative computer climate models drove the great fears of global warming that arose in the 1980s and endured through the 1990s. Early versions of those models predicted that a doubling of atmospheric CO_2 would cause global average temperature to increase by 5°C or more (nearly 10°F). As the models have been refined through the years, however, their warming predictions have moderated considerably. In 1990, the Intergovernmental Panel on Climate Change (IPCC) predicted, on the basis of the computer models, global average temperature increase of 3.3°C by A.D. 2100; by 1992, it had lowered its prediction to 2.6°C, and, by 1995, to 2.2°C (less than half the amount of warming predicted by the early computer models). Even that latest prediction is likely to turn out much too high, for it still is based on models that, had they been applied to the past century, would have predicted twice as much warming as actually occurred. As Roy W. Spencer, senior scientist at NASA's Marshall Space Flight Center, points out:

> All measurement systems agree that 1998 was the warmest year on record. The most recent satellite measurements, through 1998, give an average warming trend of +0.06°C/decade for the 20-year period 1979 through 1998. Even though this period ends with a very warm El Niño event [which would exaggerate its high-temperature end], the resulting trend is still only one-fourth of model-predicted average global warming for the next 100 years for the layer measured by the satellite.[35]

Additional uncertainties arise from significant discrepancies between temperature measurements obtained from instruments at the earth's surface and those obtained from instruments on satellites (which are substantially confirmed by instruments on weather balloons), which measure atmospheric temperature not at the surface but in the lower troposphere. These discrepancies were reported in a study prepared by the National Research

Council of the National Academy of Sciences and published in January 2000.[36] For the period 1979 through 1998, the surface data appear to indicate an average warming trend per decade of about 0.196°C (or about 1.96°C per century), while the satellite data[37] indicate a trend of only 0.057°C per decade (or about 0.57°C per century). After correcting the surface data for a variety of contaminating factors, a team of researchers produced new estimates of surface temperatures that yielded apparent decadal trends that were 0.097°C to 0.106°C larger than the satellite data trends for the lower troposphere. The differences, however, are still highly significant, since the corrected surface data trends are still 170 percent to 185 percent higher than the satellite-recorded lower troposphere trends.[38] The trouble does not end there, however. By making 1998 the final year of the study, the researchers chose a year in which global average temperatures were pushed markedly higher by an unusually strong El Niño; had the series ended with 1997 instead, the satellite data would have shown no statistically significant decadal trend, and the differential between them and the surface data would have been larger. Also, while the researchers corrected the surface data in part by accounting for the cooling effect of the eruption of Mount Pinatubo in 1991, they chose to ignore the cooling effect (about half that of Mount Pinatubo's eruption) of the eruption of Mount Chichon in 1982, further exaggerating the apparent uptrend in the satellite data.[39] The most significant problem for global warming theorists is that the computer models predicted that greenhouse warming would be faster in the lower troposphere than at the surface. But the data—to the extent that both sets are to be trusted—now show the opposite to be true. The significance of this is that the computer models clearly remain far from accurate enough in their depiction of atmospheric temperatures, which suggests that policy makers should be very slow to base their decisions on model predictions.

Not only is the actual global warming that is to be expected far from what the IPCC and other climate modelers originally predicted, but it is also questionable whether global warming is likely to bring many harmful effects. There are several reasons for this. Most important, increasingly refined models now indicate—and empirical observation has confirmed—that the majority of the warming will occur in the winter, at night, and in polar latitudes.[40] This warming is far from sufficient to cause the polar ice caps to melt, which means it is also unlikely to result in significant rises in sea level—one of the most feared results of global warming because it was thought likely to inundate many coastal cities in which millions of the world's poorest people live. Instead, nighttime warming during the winter, to the extent that it affects populated areas at all, should result in a slight decrease in energy consumption for heating (and, therefore, some reduction in future emissions) and a slight lengthening of the growing season in spring and autumn.

Further, whatever rise in global average temperature occurs will likely result not in expanding but in contracting deserts, and not in contracting but in expanding polar ice caps. Why? More water evaporates in warmer temperatures. While one might think this is bad news for deserts, the opposite is true, for deserts make up only a tiny fraction of the earth's surface; over three-fourths of it is water, and most of the remainder is moist land. But air circulates over all of it. This means that enhanced evaporation everywhere will result in enhanced rainfall, even on desert areas, which, because those areas are so dwarfed by the rest of the earth's surface, will likely receive more water by enhanced precipitation than they lose by enhanced evaporation. But the enhanced precipitation at the poles is likely to enlarge polar ice caps, offsetting a long-term natural rise in sea level. As environmental scientist S. Fred Singer points out in reviewing a variety of studies of sea level trends,

> Global sea level (SL) has undergone a rising trend for at least a century; its cause is believed to be unrelated to climate change [1]. We observe, however, that fluctuations (anomalies) from a linear SL rise show a pronounced anti-correlation with global average temperature—and even more so with tropical average sea surface temperature. We also find a suggestive correlation between negative sea-level rise anomalies and the occurrence of El Niño events. These findings suggest that—under current conditions—evaporation from the ocean with subsequent deposition on the ice caps, principally in the Antarctic, is more important in determining sea-level changes than the melting of glaciers and thermal expansion of ocean water. It also suggests that any future moderate warming, from whatever cause, will slow down the ongoing sea-level rise, rather than speed it up. Support for this conclusion comes from theoretical studies of precipitation increases [2] and from results of General Circulation Models (GCMS) [3,4]. Further support comes from the (albeit limited) record of annual ice accumulation in polar ice sheets [5].[41]

While only mild harm is to be anticipated from the small temperature increases that are most likely to come, some benefit is to be expected—indeed, has already occurred—because of enhanced atmospheric CO_2. Carbon dioxide is crucial to plant growth, and recent studies show that a doubling of atmospheric CO_2 results in an average 35 percent increase in plant growth efficiency.[42] Plants of all kinds grown in doubled-CO_2 settings become more efficient in water use, more efficient in taking up minerals from the soil, and more resistant to disease, pests, excessive heat and cold, and both floods and droughts.[43] Consequently, a portion of the great gains in agricultural productivity in the past century has been due not to intentional improvements in farming techniques but to enhanced atmospheric CO_2 caused by the burning of fossil fuels for energy to drive modern human economic activity.[44] This means that rising CO_2 has made it easier to feed the world's growing population. In addition, greater plant growth efficiency should mean—and empirical observations confirm—that plants' growth ranges will increase to higher and lower altitudes, into warmer and colder climates, and into drier and wetter climates.[45]

Some people have asserted that global warming poses a serious threat to human health through increased incidence of tropical diseases and heat-related ailments. However, the Program on Health Effects of Global Environmental Change at Johns Hopkins University, in a congressionally mandated study, "found no conclusive evidence to justify such fears"[46] but instead concluded that "the levels of uncertainty preclude any definitive statement on the direction of potential future change for each of [five categories of] health outcomes," adding, "Although we mainly addressed adverse health outcomes, we identified some positive health outcomes, notably reduced cold-weather mortality...."[47] As the report exemplifies, it is easy for researchers to focus only on anticipated negative health effects from changes in global atmospheric chemistry and climate. However, not only must such anticipated effects be carefully justified and quantified in themselves, but they must also be studied in balance with anticipated benefits. For example, the reduction in hunger and malnutrition attributable to rising agricultural yields from increased atmospheric carbon dioxide, however difficult to quantify, must certainly be considered. Thomas Gale Moore concluded his careful evaluation of various studies of anticipated health effects of global warming by writing, "... a warmer climate should improve health and extend life, at least for Americans and probably for Europeans, the Japanese, and people living in high latitudes. High death rates in the tropics appear to be more a function of poverty than of climate. Thus global warming is likely to prove positive for human health."[48] What is clear is the need for added study before long-term, difficult-to-change policies are adopted.

Despite all this, some people still want to greatly curtail fossil fuel use to reduce CO_2 emissions. They are promoting a number of measures to do so, such as the Kyoto Protocol, an international treaty to force reductions in energy consumption. But since every form of economic production requires energy, reducing energy use entails reducing economic production. Some will reply that the losses in production can be offset by improved energy efficiency. To some extent they might be, but it is very unlikely that the reductions in emissions could be achieved through government-mandated efficiency measures alone; almost certainly, some actual loss of production would result. Because individuals seek to reduce their cost of living and businesses seek to maximize their profits in a free and competitive economy, they have a natural incentive to minimize waste, that is, to eliminate inefficient behavior and adopt the most economically efficient technologies they can (though these are not always the most technically efficient). The apparent need for government to mandate further emission reductions therefore suggests that these reductions must cause a net loss in production and, ultimately, diminish human welfare.

The independent economic forecasting firm WEFA, even after accounting for likely improvements in energy efficiency, estimates that meeting the United States targets under the Kyoto accords would cut annual economic output by about $300 billion (or about 3.5 percent of the roughly $8.4 trillion in 1998 gross domestic product [GDP]) and, by 2010, destroy more than 2.4 million jobs and reduce average annual family income by about $2,700. Another economic forecasting firm, Charles River Associates, projects lower costs—about 2.3 percent (or, currently, about $193 billion) of GDP per year. Whether higher or lower, these economic costs translate into very human costs. Specialists in risk assessment estimate that in the United States, every $5 to $10 million drop in economic output results in one additional statistical death per year.[49] At that rate, the loss of $193 to $300 billion in annual economic output entails at least 19,300 to 30,000 additional premature deaths per year in the United States alone.

But the United States is a rich country, far better able to cope with the costs of Kyoto than the vast majority of the world. The lost economic growth in any developing countries that are forced to comply with Kyoto emission restrictions spells added decades of suffering and premature deaths for their people, for whom the affordability of basic water and sewage sanitation, health care, and safe transportation will be long postponed.

Thus, says Frederic Seitz, past president of the National Academy of Sciences, in a letter accompanying a petition against the treaty signed by over seventeen thousand scientists,[50]

> This treaty is, in our opinion, based upon flawed ideas. Research data on climate change do not show that human use of hydrocarbons is harmful. To the contrary, there is good evidence that increased atmospheric carbon dioxide is environmentally helpful. The proposed agreement would have very negative effects upon the technology of nations throughout the world, especially those that are currently attempting to lift from poverty and provide opportunities to the over 4 billion people in technologically underdeveloped countries.[51]

Even assuming that the popular global warming scenario were true, what benefit would come from all the costs—not just in the United States but all over the world—of complying with the Kyoto accords? Proponents of the accords estimate that without the Kyoto limits, hydrocarbon emissions will increase at about 0.7 percent per year and that this will raise effective atmospheric carbon dioxide concentration from the present level of about 470 parts per million (PPM) to about 655 PPM in the year 2047. The Kyoto Protocol calls for reduction of emissions to 7 percent below 1990 levels during the years 2008 to 2012 and no increase thereafter, with effective carbon dioxide concentration in 2047 of 602 PPM. How much warming would be prevented by then? About 0.19°C out of a potential 0.5°C.[52] At a cost to the United States alone of about $200 billion per year (slightly above the Charles River Associates estimate but only two-thirds of the WEFA estimate), this would mean a total cost of roughly ten trillion dollars and one million premature deaths. Such a price is too much to pay for so small and doubtful a benefit.

Not only the highly uncertain nature of both the theory and the evidence of global warming but also the unresolved question of whether global warming's net effects will be negative or positive point to one sure policy for the present: to delay action—especially highly costly action such as mandatory reductions in energy consumption—until the matter is much better understood.

It is tempting to say that we must not politicize this (or any other) environmental issue, and we do not intend to do so; our focus is on sound science rooted in a value structure that emphasizes honesty and openness to debate and evidence. But the issue has already been heavily politicized. Starting in the early 1990s, advocates of the Kyoto Protocol frequently spoke of a "scientific consensus" about global warming and derided the motives of scientists and others who questioned that conclusion. More recently, Rev. Dr. Joan Brown Campbell, general secretary of the National Council of Churches, went so far as to say that belief in global warming and support for the Kyoto Protocol should be "a litmus test for the faith community."[53] Clearly, as a result of such thinking, the quality of public knowledge and, hence, the ability to make wise public policy decisions, have been badly compromised with regard to global warming. Massachusetts Institute of Technology meteorology professor Richard Lindzen, one of the leading researchers in greenhouse effect and climate change science, pointed out in the early 1990s that "the existence of large cadres of professional planners looking for work, the existence of advocacy groups looking for profitable causes, the existence of agendas in search of saleable rationales, and the ability of many industries to profit from regulation, coupled with an effective neutralization of opposition" have undermined the quality of debate over both science and public policy, and that

> the dangers and costs of those economic and social consequences may be far greater than the original environmental danger. That becomes especially true when the benefits of additional knowledge are rejected and when it is forgotten that improved technology and increased societal wealth are what allow society to deal with environmental threats most effectively. The control of societal instability [brought on by the politicization of science in the global warming debate] may very well be the real challenge facing us.[54]

Contrary to earlier claims, it turned out that there was no consensus in favor of the popular global warming scenario. Even in the early 1990s, when the National Research Council appointed a panel dominated by environmental advocates—a panel that included Stephen Schneider, who is an ardent proponent of the catastrophic hypothesis—the panel concluded that there was no scientific basis for any costly action.[55] If any scientific consensus has grown since then, it has been critical of the catastrophic vision and the policies based on it. First, like a warning shot across the bow, came the Statement by Atmospheric Scientists on Greenhouse Warming, released February 27, 1992. Signed by forty-seven atmospheric scientists, many of whom specialized in global climate studies, it warned that plans to promote a carbon emissions reduction treaty to fight global warming at the upcoming Earth Summit in Rio de Janeiro in June 1992 were "based on the unsupported assumption that catastrophic global warming follows from the burning of fossil fuels and requires immediate action," adding, "We do not agree." It cited a 1992 survey of United States atmospheric scientists, conducted by the Gallup organization, demonstrating that "there is no consensus about the cause of the slight warming observed during the past century." Further, the statement cited "a recently published paper [that] suggests that sunspot variability, rather than a rise in greenhouse gases, is responsible for the global temperature increases and decreases recorded since about 1880." It continued, "Furthermore, the majority of scientific participants in the [Gallup] survey agreed that the theoretical climate models used to predict a future warming cannot be relied upon and are not validated by the existing climate record," and it pointed out that "agriculturalists generally agree that any increase in carbon dioxide levels from fossil fuel burning has beneficial effects on most crops and on world food supply."[56] This was followed by the Heidelberg Appeal, released at the Earth Summit. Although it did not specifically name global warming, the Heidelberg Appeal warned against "the emergence of an irrational ideology which is opposed to scientific and industrial progress and impedes economic and social development." Over three thousand scientists, including seventy-two Nobel Prize winners, signed it.[57]

Three years later came the Leipzig Declaration on Global Climate Change, developed at the International Symposium on the Greenhouse Controversy held in Leipzig, Germany, in November 1995, and revised and updated after a second symposium there in November 1997. Signed by eighty leading scientists in the field of global climate research and twenty-five meteorologists, the document declared "the scientific basis of the 1992 Global Climate Treaty to be flawed and its goal to be unrealistic," saying it was "based solely on unproven scientific theories, imperfect climate models—and the unsupported assumption that catastrophic global warming follows from an increase in greenhouse gases." It added, "As the debate unfolds, it has become increasingly clear that—contrary to conventional wisdom—there does not exist today a general scientific consensus about the importance of greenhouse warming from rising levels of carbon dioxide.

In fact, most climate specialists now agree that actual observations from both satellite and balloon-borne radiosondes show no current warming whatsoever—in direct contradiction to computer model results." And it concluded, "based on all the evidence available to us, we cannot subscribe to the politically inspired world view that envisages climate catastrophes and calls for hasty actions. For this reason, we consider the drastic emission control policies deriving from the Kyoto conference—lacking credible support from the underlying science—to be ill-advised and premature."[58]

But those early signs of consensus against the popular vision were dwarfed by the release in 1997 of a Global Warming Petition developed by the Oregon Institute of Science and Medicine and accompanied by a thoroughly documented review monograph on global warming science. The petition urged the rejection of the Kyoto Protocol "and any other similar proposals," saying boldly, "The proposed limits on greenhouse gases would harm the environment, hinder the advance of science and technology, and damage the health and welfare of mankind." It added,

> There is no convincing evidence that human release of carbon dioxide, methane, or other greenhouse gases is causing or will, in the foreseeable future, cause catastrophic heating of the Earth's atmosphere and disruption of the Earth's climate. Moreover, there is substantial scientific evidence that increases in atmospheric carbon dioxide produce many beneficial effects upon the natural plant and animal environments of the Earth.[59]

The Global Warming Petition was signed by more than 17,000 basic and applied American scientists, including over 2,500 physicists, geophysicists, climatologists, meteorologists, oceanographers, and environmental scientists well qualified to evaluate the effects of carbon dioxide on the earth's atmosphere and climate, and over 5,000 chemists, biochemists, biologists, and other life scientists well qualified to evaluate the effects of carbon dioxide on plant and animal life. The consensus of scientists on global warming has turned out to be quite the opposite of what the apocalyptic vision proponents claimed.

Species Extinction

The Bible clearly indicates that God takes delight in his many creatures (Job 38:39–39:30; 40:15–41:34; Ps. 104:14–23). This entails the importance of stewardship of life itself. Confronted with claims that anywhere from 1,000 to 100,000 species are going extinct per year and that many or most of the extinction is caused by human action,[60] Christians must wonder whether they have failed in their stewardship obligation. However, in the spirit of 1 Thessalonians 5:21 ("Test all things; hold fast to what is good"), we can insist that claims of species extinction rates be tested empirically and that the significance of these numbers be carefully evaluated in the proper context.

When the claims are tested, they are found to be highly dubious. When two eminent statisticians challenged the claims, asserting that no empirical field data existed to support them,[61] the International Union for the Conservation of Nature (IUCN) responded by commissioning a major worldwide field study. The result was a book[62] in which author after author admits that, despite expectations to the contrary based on theoretical models, field research yields little or no evidence of species extinction, even in locales—such as heavily depleted rain forests—in which the highest rates were anticipated. In that volume, V. H. Heywood, former director of the scientific team that produced the Flora Europea, the definitive taxonomic compilation of European plants, and S. N. Stuart, executive officer of the Species Survival Commission at the IUCN, wrote, "IUCN, together with the World Conservation Monitoring Centre, has amassed large volumes of data from specialists around the world relating to species decline [worldwide], and it would seem sensible to compare these more empirical data with the global extinction estimates. In fact, these and other data indicate that the number of recorded extinctions for both plants and animals is very small." They add,

> Known extinction rates [worldwide] are very low. Reasonably good data exist only for mammals and birds, and the current rate of extinction is about one species per year.... If other taxa were to exhibit the same liability to extinction as mammals and birds (as some authors suggest, although others would dispute this), then, if the total number of species in the world is, say, 30 million, the annual rate of extinction would be some 2,300 species per year. This is a very significant and disturbing number, but it is much less than most estimates given over the last decade.[63]

Note, however, that this hypothesis of 2,300 extinctions per year is not based on empirical evidence; it is instead derived from a theoretical model of extinctions as a percentage of total species and a high guess of total species. A more likely estimate of total species might be five to ten million, which, inserted into the model, would yield about 380 to 770 extinctions per year. If those numbers still sound alarming, keep in mind, first, that they represent only about 0.008 percent of species per year and, second, that they are probably significantly exaggerated. Even at that rate, it would take over five hundred years to eliminate 4 percent of all species on earth. What is more, as already noted, the same book contains repeated admissions that the model predictions of high extinction rates were repeatedly falsified by field investigation.

That is not surprising to those familiar with the serious weaknesses in the species-area curve and island biogeography theories from which the hypothetical extinction rates are derived. Subjected to careful critique, they turn out to vastly overestimate real extinction rates. In part, this is because they fail to describe ecosystems as they really are, and they unrealistically attribute to large, connected regions (e.g., the Amazon rain forest) the characteristics of isolated islands.[64] This means it is likely that the real extinction rate is much lower than 0.008 percent of species lost per year.

In short, the lack of sound data to support claims of species extinction rates continues.[65] Instead, the observational data indicate very low rates of extinction. A World Conservation Union report in 1994 found extinctions since 1600 to include 258 animal species, 368 insect species, and 384 vascular plants—about 2.5 species lost per year.[66] Consider the loss of species in the United States:

> Of the first group of species listed in 1973 under the Endangered Species Act, today [1995] 44 are stable or improving, 20 are in decline, and only seven, including the ivory-billed woodpecker and dusky seaside sparrow, are gone. This adds up to seven species lost over 20 years from the very group considered most sharply imperiled.... Under [conservation biologist E. O.] Wilson's loss estimate of 137 species per day, about 1.1 million extinctions should have occurred globally since 1973. As America contains six percent of the world's landmass, a rough proration would assign six percent of that loss, or 60,000 extinctions, to the United States. Yet in the period only seven actual U.S. extinctions have been logged.... And the United States is the most carefully studied biosphere in the world, making U.S. extinctions likely to be detected.

If plants and insects are included in the calculation, 34 organisms fell extinct in the United States during the 1980s, according to a study by the Department of the Interior. This is clearly worrisome, but at an average of 3.4 extinctions per year, nothing like the rate of loss claimed by pessimists.[67]

The significance even of these small numbers is open to debate because, while most people think of a species as genetically defined, the Endangered Species Act (ESA) defines species very differently. The Act says, "The term 'species' includes any subspecies of fish or wildlife or plant, and *any distinct population segment* of any species of vertebrate fish or wildlife which interbreeds when mature" (emphasis added).[68] The trouble with this definition is that when most people unfamiliar with the ESA think of a species as being in danger of becoming extinct, they think this means no individual organism of that genetic definition will be left anywhere—or, since the ESA applies to the United States, at least there. (This popular perception certainly lies behind the fear that "species" extinction forever removes elements from the global gene pool.) But in reality, it may only mean that a given population segment of that genetically defined species is endangered; it is entirely possible that plenty of other specimens may thrive in other locations. Many citizens who support expensive policies to prevent species extinctions might reconsider if they knew that rather than preventing real extinctions, they were only preventing the removal of a geographically defined segment of an otherwise thriving species.

None of this means that there are not particular species that are, in fact, endangered and that can benefit from careful conservation efforts. But as field ecologist Rowan B. Martin points out, when monetary values are more fully aligned with other human values, the institutional arrangement allows for the maximization of both values:

> Western scientists, activists, and agencies favor the creation of reserves in developing nations to preserve biological diversity. However, this strategy is often an unworkable form of "eco-imperialism." Recent studies show that the majority of reserves are failing to conserve biodiversity, are financially unsustainable, and were irrelevant to 95 percent of the people in the countries where they were located. An alternative strategy, which has had considerable success, is empowering local people to control the wildlife

resources in their area. In many parts of Southern Africa, where full rights of access and control over wildlife have been granted to landholders (of both private and communal land), biodiversity is better conserved in the areas surrounding national parks than in the parks themselves. Additionally, the areas surrounding the parks are economically more productive than the state-protected areas. In Southern Africa and other parts of the world, conservation of biological resources would be a profitable activity and not a cost if the correct institutional arrangements were developed, including a stronger reliance on private property and communal tenure systems.[69]

V. Environmental Market Virtues[70]

We have already argued that economic growth itself is an important step toward environmental protection. It makes good stewardship affordable and technically possible. Nonetheless, economic growth by itself is not enough. Human initiative needs to take place within an institutional framework that promotes environmental stewardship. Therefore, we need to examine more closely what is institutionally necessary to help further the goal of environmental protection.

While some concerns about the environment are overstated, others are quite real and need our attention. The fact that the world is not experiencing overpopulation or destructive, manmade global warming or rampant species loss does not mean that a change in policies or practices is not needed to address other issues.

Christians have every reason to embrace an appropriate environmental ethic, one that honors creation but distinguishes it from the Creator. However, simply recommending reformation of our worldview is not sufficient. Our ability to act responsibly toward nature has been hindered by our alienation from God. The original Fall and our continued rebellion mean that we act selfishly, that we have limited knowledge, and that we often fail to recognize the full potential in the created order. In view of these failings, we must not rely on worldview alone to lead us to good decisions about creation but must also examine the other influences of decision making, namely, information and incentives.

Environmental problems are traditionally seen as a result of market failure and as ample justification for the government to involve itself in the economy much more directly and forcefully to solve these problems. But it is an error to assume that, just because the market does not presently solve certain problems, government can effectively intercede to do so. Information and incentives are very much affected by the institutional order of a society. The social institutions pertinent to environmental and resource issues are the rules that assign responsibility—that is, property rights that determine who can take what actions and who gets a hearing with regard to those actions. These rules are crucial determinants of what information is generated and what incentives the decision makers face.

Property rights generate appropriate information and incentives to the extent that they embody three characteristics: exclusivity, liability, and transferability. Exclusivity means that the owner of a resource is able to capture a return from using the property in a way that is advantageous to other people, and it also means that an owner can exclude others from benefiting from the use of the property unless they have secured the owner's permission. If exclusivity does not exist, a resource will be overused. For instance, on the American frontier there were no exclusive rights to North American buffalo. If a buffalo hunter decided to postpone the shooting of any particular animal, he had no assurance that he would have the option to exercise that right in the future. The only way he could be assured of an exclusive right to a buffalo was to shoot it. Live buffalo were owned by everyone; dead ones belonged to the person who killed them. Is it any wonder that such a property rights system led to the near-extinction of the species?[71]

Liability forces a resource owner to bear the costs of actions that harm others. If property rights fully embody liability, costs are not imposed on others without their willing consent. For instance, if a person allows another person to impose harm on him—that is, to use up some of the grass on his cattle ranch to feed his livestock—that person must receive what he believes to be adequate compensation for the harm. If liability were not fully attached to one's property—that is, one's cattle—a person could drive cattle across someone else's land, allowing them to remove some of the grass without providing compensation. Pollution is a notable example of an incomplete property right, of liability not being present. It is exactly analogous to the cattle example; individuals can use up some of another's resource—clean air—without appropriate compensation.

Transferability encourages owners to look for ways of using property that benefit others, a central obligation of the Christian faith. The fact that a piece of property can be bought or sold means that a resource owner who ignores the wishes of other people does so at a cost to himself, a reduction of wealth. If rights are not transferable, no such wealth loss is associated with ignoring the wishes of others. In other words, transferability encourages people to seek out and engage in the most mutually beneficial property arrangements possible.

Thus, the attributes of exclusivity, liability, and transferability are essential for a well-functioning property rights system, one that fulfills the biblical mandate of holding individuals accountable for their decisions. If any one of those attributes is missing, people can act irresponsibly with regard to creation, at least in part because they do not have adequate information or appropriate incentives to make sound decisions.[72]

The information available to a decision maker is very much a function of property rights because people, in the process of trading, generate indexes of value for various uses of property. For instance, a landowner who knows there is coal on his land can readily obtain information through the price system about how others in society value that coal. If that individual also holds rights to the coal, that same information contains incentives for the owner to take actions that satisfy other people, namely, to make coal available to them. Since part of the biblical mandate with regard to creation is to use it for humankind, it would seem to be appropriate to be aware of and respond to people who desire to use coal as a fuel source.

But is mining the coal the only use for that land? What if mining leaves ugly scars on the earth's surface, permanently reducing certain individuals' aesthetic enjoyment of that land? How does a price system take those desires into account? Will coal be mined while aesthetics are ignored? The price system does not adequately represent all desires, and its failure to do so is caused by a lack of appropriate property rights. If the landowner had exclusive control over view rights to her land, she could charge an appropriate fee, and the price system would communicate to her whether the land was more valuable left in its pristine state or mined for coal.

The fact that property rights are sometimes not well defined and enforced is at the heart of environmental despoilment. The lack of a full rights structure means decision makers do not have appropriate incentives and information. Therefore, it is not surprising that resource misuse occurs when property rights are incomplete. Of course, simply pointing out the lack of adequate property rights is not a solution to the environmental problem, but it provides some general guidance. We do not necessarily want to fully define rights to all resources; in some cases, the transaction costs of doing so are too high. But many property rights problems are not intractable, and the property rights framework is a useful way of looking at environmental issues.

For instance, air and water are the major resources suffering from pollution in certain places because they are usually treated as common property, that is, property where no one has exclusivity. Any individual who uses a particular airshed or watershed to dispose of waste does not face the full cost of his action; instead, the costs are spread over all the potential users of that resource, resulting in what has been called the "tragedy of the commons."[73] The answer to this problem is to attempt to restructure property rights so that exclusivity, liability, and transferability exist. Sometimes there are legal barriers to property rights' definition and transfer, as in the case of water law in many states, and those barriers can be removed. In others, the government must take positive steps to force decision makers to bear the full costs of their actions. For instance, a tax per unit of air or water pollution increases the costs of using the air or water as a waste disposal mechanism. If the tax is set at the correct level (if it accurately represents the cost of pollution—a difficult proposition when set outside of a market framework), the decision maker faces the correct incentive structure. He can continue to pollute if he is willing to pay the cost, and, if he does, the additional benefits to society from the polluting activity exceed the additional costs. In all likelihood, under such a tax the polluter will decide to reduce emissions—but not to zero.

Another way of altering property rights in air is through "the bubble concept." Under such a structure, people residing in a particular airshed, through some government entity, would decide how much pollution they are willing to tolerate. Rights to the pollution would then be available to producers in the area. The rights could be either handed out on the basis of historical production or auctioned off to the highest bidder. An important element of such a system would be transferability; for the rights to result in the greatest production at the lowest cost, each pollution right would need to be fully transferable within the airshed. Then each producer would face an appropriate incentive structure and could decide if it would be cheaper to purchase pollution rights and continue polluting at the company's historical rate, or to adopt pollutant-reducing technology, or to shut down.

Each of these proposals involves government action of some sort. Because the definition and the enforcement of property rights are at least, in part, a function of government, an alteration of those rights will probably involve government. However, one must carefully specify the type of action appropriate when suggesting that government is the answer to environmental problems. Seeing the problem as one of inadequate property rights gives positive guidance about how government can be most effective—through the clear specification of rights and the fuller defense of them. Unfortunately, too often, government's involvement in resource issues has not been framed in a property rights context and hence has not been as effective as possible.

For instance, in terms of air and water pollution, the common governmental response has been through a command-and-control approach. Under such a system, government specifies the amount of pollution that can occur from each source and, in many cases, also specifies the technology to be used in reducing emissions. Numerous studies have shown that for any goal achieved through command-and-control, a bubble concept with transferable rights could achieve the same level of pollution reduction much more cheaply. [74]

The oft-repeated suggestion that government ownership and management of resources are solutions to environmental problems might seem to be appropriate when private property rights and markets have failed to lead to sound resource management. However, this suggestion ignores the fact that under government ownership, it is very difficult to construct property rights so that decision makers face appropriate incentives and receive correct information.

An excellent example of how governmental attempts at stewardship can create perverse incentives involves the Endangered Species Act (ESA). This legislation, rather than creating incentives for people to act as good stewards of their own land and of its plant and animal inhabitants, often has exactly the opposite effect by making people fearful of losing use of that land. Richard Stroup, one of the originators of the New Resource Economics, describes the incentives of the ESA in this way:

> Under the Endangered Species Act, the owner must sacrifice any use of the property that federal agents believe might impair the habitat of the species—at the owner's expense. Furthermore, if the owner either harms the species or impairs its habitat, severe

penalties are imposed. The perverse incentives created by the law may well lead an owner to surreptitiously destroy that animal or plant—or any habitat that might attract it.[75]

Utah State University political science professor Randy Simmons observes that "the Supreme Court declared in its Tellico Dam decision that the act defines 'the value of endangered species as incalculable,' that endangered species must 'be afforded the highest of priority,' and that 'whatever the cost' species loss must be stopped (*TVA v. Hill*, 437 U.S. 187, 174, 184 [1978])."[76] Such a zealous legislative commitment ignores the full scale of human values that a free economy otherwise allows to show through in the pricing system. But such a commitment by government turns the real value of a species from an asset into a liability—for instance, from the satisfaction one feels from having a rare species live on one's land to the fear of losing the use of land essential to one's livelihood. As field ecologist Rowan Martin argued earlier about wildlife resource preserves in southern Africa, empirical observation confirms that, when monetary values are more fully aligned with other (such as environmental) values, the institutional arrangement allows for the maximization of both values.

How do we know that the desires represented through property rights and the markets are truly scriptural? Is it not possible to have a well-functioning market system and still have resources put to ungodly uses? At this point, the biblical environmental ethic must inform the private-property system. An institutional structure that embodies exclusivity, liability, and transferability in its property rights will accurately represent the desires of members of society and will also encourage resource owners to respond to those desires. Full accountability—a biblical concept—will be in place. However, one must remember that Scripture most often discusses accountability in the context of responsibility to God, and the accountability being discussed here is accountability to other people, which is an entirely different concept.

All of this reaffirms the need for a biblically based view of nature and of man so that the desires represented in the marketplace will come closer to God's desires. At the same time, however, it is not clear that any alternative democratic institutional structure would lead to a more godly environmental policy. The biblical mandate of valuing nature but making use of it does not offer much guidance as to the particulars of resource use.

Evidently, God has allowed man to work out those details on the basis of his own perceptions of needs—with those needs appropriately informed by an awareness of God and his principles.

We are limited by human desires, as imperfect as they might be, as our standard to measure how resources should be used. God has given us the opportunity and responsibility to manage his creation, and it therefore seems appropriate to have an institutional structure that reflects human desires and holds individuals accountable as to whether they use their resources according to those desires. Such a structure is the system of property rights described earlier. If this seems a weak defense of property rights, that may be because it is. One can conceive of many cases where a system of well-defined and enforced property rights results in resource use that seems to violate God's standards. However, it is difficult to conceive of another property rights structure that does better at making sure God's standards are not violated. The two most obvious alternatives—common property and government ownership—both suffer from such obvious faults, such as the tragedy of the commons, that they are clearly inferior choices.

Despite this rather lukewarm endorsement of private-property rights as the correct mechanism for controlling resource use, several facets of such a system deserve some approbation. Such a rights structure allows for expression of certain aspects of the biblical principles outlined in the first section of this paper.

First, a private-property system will not produce zero pollution in the sense of stopping all alteration of the environment; but neither will it allow economic growth at all costs with material desires superseding all others. If property rights are fully defined and enforced, some emissions will still foul our air, not all water will be of pristine quality, and the use of nonrenewable resources will not drop to zero. However, the significant difference between this potential system of private-property rights and the the one that currently exists is that actions altering the environment would take place only if all users of the environment were convinced that those actions were to everybody's mutual advantage. In other words, there would be no uncompensated losers. A person who valued an unspoiled view more than someone else valued a factory smokestack in the middle of that view would win out. The factory smokestack would not exist, at least not at that location. Such a property rights system would not stop economic growth but would allow it to occur only if the benefits were valued more highly than what was given up to get that growth. Such an approach to resource use seems appropriate, as we are to appreciate and value God's creation, but also see it as usable for human purposes.

Another component of a private-property rights system is that it does not depend on complete social agreement for action to take place. Diversity is permitted by virtue of the fact that a person who has strong feelings about resource use that differ from the group consensus can, under such a system, express those feelings through prices and markets. This can be of particular importance to Christians or environmentalists who find themselves at odds with prevailing wisdom about the environment. If such beliefs represent a minority position, they are much more likely to find expression in a system of private-property rights than under alternative rights arrangements.

Finally, a private-property rights system permits the fullest realization of the image of God in the human person. Genuine problems require genuinely creative solutions, and property harnesses human creativity to the realization of human needs. As history has repeatedly shown, it is the creative spirit of the human person that permits wise stewardship, and institutions that encourage this spirit are more likely to also facilitate environmentally sound ends.

But can we be assured that future generations will have a place in a free economy? What of God's concern for all people of all times? Is there not a chance that a system based on private-property rights will cater exclusively to the desires of the present generation compared to the needs of future ones? Again, the appropriate question to ask is, Compared to what? What alternative institutional arrangement will do a better job than one that embodies transferable property rights? It would be nice to posit a theocracy headed by an omniscient saint, and if that were a realistic alternative, markets would come out second-best. However, if we stick to real-world possibilities, well-defined rights that can be bought and sold look quite good indeed.

Contrast, for a moment, a resource being managed under two alternative regimes. Let us say that a resource is exhaustible; hence, it is important to give future generations some voice in the choice about the appropriate rate of use. Under the first regime, a pure democracy controls the use of the resource. With different expectations by members of the population about the resource's future value, the average perception will dominate. In other words, if the present generation thinks that, on average, the resource has a future value (discounted to the present) greater than its value in present consumption, it will be preserved. On the other hand, if the average expectation of the resource's future value is less than its value in present consumption, it will be consumed.

Now take the same resource, and the same population with the same set of preferences and expectations, but make the present/future allocation on the basis of transferable property rights. In this case, the resource is more likely to be preserved for the future because it is not the average perception about the future value of the resource that counts, but instead the perception of those most optimistic about its future value who express themselves in the marketplace. These individuals will purchase the resource in the expectation of a high future value, hold it out of consumption, and, in the process, preserve it for future generations. In fact, for any resource to be used in the present, all who believe it has some value in the future must be outbid.

All of this is not to say that altruistic feelings for future generations are unimportant. Under either system, such sentiments can result in greater preservation for future generations. Notice, however, that the political approach depends entirely on altruism, or people caring for future generations, while the market order allows those preferences to be expressed but also rewards individuals who, for selfish reasons, decide to withhold resources from present consumption.

Giving future generations a voice is a bit awkward. Their preferences will be expressed only in people who exist presently, so it is useful to have someone stand in for them today; they need agents to represent them. These agents cannot know perfectly the desires of people not yet born, but they can make educated guesses about these desires. In the market arena, these agents are either unselfish contributors to the future or speculators acting on their perception of future demands for resources. If their perceptions are correct, their wealth increases; if they guess incorrectly, they suffer a wealth loss. Thus, these agents have strong incentives to be well informed and to predict correctly the needs of future generations.

In a world where Christian charity and concern for others are sometimes in short supply, it is useful to have a mechanism that allows for future needs to be met, by those acting charitably and those pursuing profit. Again, institutional design is a fundamental component of a system that satisfies God's desire that we think not only of this generation.

Thus, freedom, property rights, and a legal framework that ensures that accountability attaches to freedom and property, work together to minimize pollution and improve human welfare. As Carl Pope, president of the Sierra Club, has noted, this sort of approach "would yield restrictions on pollution more stringent than those embodied in any current federal and state pollution laws,"[77] without necessarily sacrificing human welfare in the process.

The more fully, then, a society embodies a Christian worldview, and the more its decision makers—private and public—embrace that value framework and operate with the information and incentives provided by a private-property legal regime with exclusivity, liability, and transferability, the more decisions with environmental impact are likely to be responsible and to minimize harm to people and the larger environment. The Christian worldview can be promoted by preaching, teaching, writing, and the like. But the information and incentives essential to proper decision making, even assuming a Christian worldview, are best generated by the price system of the free economy.

Conclusion

Patrick Moore, one of the founders of Greenpeace International, said in an interview in the *New Scientist* in December 1999, "The environmental movement abandoned science and logic somewhere in the mid-1980s ... political activists were using environmental rhetoric to cover up agendas that had more to do with class warfare and anti-corporatism than with the actual science...." What we have said above indicates that Moore was right in his critique of the movement to which he made such an important early contribution. Too often, modern environmentalism has become anti-human, anti-freedom, anti-economic development, and anti-reason. It is time to reverse this trend.

On the basis of a biblical worldview and ethics, as well as of sound science, economics, and public policy principles, we believe sound environmental stewardship celebrates and promotes human life, freedom, and economic development as compatible with, even essential for, the good of the whole environment. While we do not rule out all collective action, we believe market mechanisms are frequently better means, in both principle and practice, to environmental protection. They are less likely to erode important human freedoms and more likely to be cost-effective and successful in achieving their aims. While we understand that passions may energize in the pursuit of sound environmental policy, we also believe that reason, coupled with a commitment to "do justly, to love mercy, and to walk humbly with … God" (Mic. 6:8), must ultimately guide environmental policy.

Editorial Board

E. Calvin Beisner, Associate Professor of Historical Theology and Social Ethics, Knox Theological Seminary, and Adjunct Fellow, Committee for a Constructive Tomorrow

Michael Cromartie, Vice President and Director of Evangelical Studies, Ethics and Public Policy Center

Dr. Thomas Sieger Derr, Professor of Religion, Smith College

Dr. Peter J. Hill, President, Association of Christian Economists, and Professor of Economics, Wheaton College

Diane Knippers, President, Institute for Religion and Democracy

Dr. Timothy Terrell, Professor of Economics, Liberty University

Notes

1. Robert William Fogel, "The Contribution of Improved Nutrition to the Decline in Mortality Rates in Europe and America," in *The State of Humanity*, ed. Julian L. Simon (New York: Blackwell, 1995), 61–71.

2. E. Calvin Beisner, "Sixpence None the Richer: Economics—A Millennium of Human Progress," *World* 14 (July 31, 1999): 20–25. For voluminous statistics and able discussions on these and dozens of other elements of material progress, see Julian L. Simon, ed., *The State of Humanity* (New York: Blackwell, 1995).

3. See E. Calvin Beisner, *Prosperity and Poverty: The Compassionate Use of Resources in a World of Scarcity* (Wheaton, Ill.: Crossway Books, 1988), and *Prospects for Growth: A Biblical View of Population, Resources, and the Future* (Wheaton, Ill.: Crossway Books, 1990); and Nathan Rosenberg and L. E. Birdzell, Jr., *How the West Grew Rich: The Economic Transformation of the Industrial World* (New York: Basic Books, 1986).

4. Nicholas Eberstadt, "World Depopulation: Last One Out Turn Off the Lights," *Milken Institute Review* 2 (first quarter 2000): 38.

5. The classic work leading to biological egalitarianism is Peter Singer's *Animal Liberation: A New Ethics for Our Treatment of Animals* (New York:

Random House/New York Review of Books, 1975). See also John Harris, Stanley Godlovitch, and Roslind Godlovitch, *Animals, Men, and Morals* (New York: Taplinger Publishing, 1972); and Arne Naess, *Ecology, Community, and Lifestyle: Outline of an Ecosophy*, trans. and rev. David Rothenberg (Cambridge and New York: Cambridge University Press, 1989). For critique, see E. Calvin Beisner, *Where Garden Meets Wilderness: Evangelical Entry into the Environmental Debate* (Grand Rapids, Mich.: Eerdmans Publishing/Acton Institute, 1997), appendix 2; Thomas Sieger Derr, *Environmental Ethics and Christian Humanism* (Nashville, Tenn.: Abingdon Press, 1996), chapter 1, and "Human Rights and the Rights of Nature," *Journal of Markets and Morality* (forthcoming); Robert Royal, *The Virgin and the Dynamo: Use and Abuse of Religion in Environmental Debates* (Grand Rapids, Mich.: Eerdmans Publishing, 1999), chapter 4; and Charles T. Rubin, *The Green Crusade: Rethinking the Roots of Environmentalism* (New York: Free Press, 1994), chapter 4.

6. Quoted in Francis A. Schaeffer, "How Should We Then Live?" in *The Complete Works of Francis A. Schaeffer: A Christian Worldview* (Westchester, Ill.: Crossway Books, 1982), 5:159.

7. See Michael B. Barkey, "A Framework for Translating Environmental Ethics into Public Policy," *Journal of Markets and Morality* (forthcoming); E. Calvin Beisner, "Stewardship in a Free Market," in *The Christian Vision: Morality and the Marketplace*, ed. Michael Bauman et al. (Hillsdale, Mich.: Hillsdale College Press, 1994), and *Where Garden Meets Wilderness: Evangelical Entry into the Environmental Debate* (Grand Rapids, Mich.: Eerdmans Publishing/ Acton Institute, 1997), appendix 2; Thomas Sieger Derr, *Environmental Ethics and Christian Humanism* (Nashville, Tenn.: Abingdon Press, 1996), chapter 1, and "Human Rights and the Rights of Nature," *Journal of Markets and Morality* (forthcoming); and Peter J. Hill, "Biblical Principles Applied to a Natural Resources/Environment Policy," in *Biblical Principles and Public Policy: The Practice*, ed. Richard Chewning (Colorado Springs: NavPress, 1991), 169–182.

8. Scripture frequently defines justice procedurally as rendering impartially and proportionally to everyone his due in accord with the standards of God's moral law. Elements of this definition are found throughout Scripture: impartiality (Lev. 19:15; Deut. 16:19; 1 Tim. 5:21; James 2:1–9); moral desert (Prov. 24:12, cf. Matt. 16:27; Rom. 2:6; 13:7; 1 Cor. 3:8; Gal. 6:7–8); proportionality (Exod. 21:35–36; 22:1, 6; Lev. 24:17–21; Deut. 19:4–6); and conformity to a standard (Lev. 19:35–37; Deut. 25:13–16, cf. Job 31:6, Ezek. 45:10, and Mic. 6:8.). For a discussion of recent debates among evangelicals over the meaning and nature of justice and the implications this has for political economy, see Craig M. Gay, *With Liberty and Justice for Whom? The Recent Evangelical Debate over Capitalism* (Grand Rapids, Mich.: Eerdmans Publishing, 1991).

9. James Gwartney and Robert Lawson, with Dexter Samida, *Economic Freedom of the World, 2000 Annual Report* (Vancouver: Fraser Institute, 2000), 15.

10. See, for example, Indur M. Goklany, "Richer Is Cleaner: Long-Term Trends in Global Air Quality," in *The True State of the Planet*, ed. Ronald Bailey (New York: Free Press, 1995), and "Richer Is More Resilient: Dealing with Climate Change and More Urgent Environmental Problems," in *Earth Report 2000: Revisiting the True State of the Planet*, ed. Ronald Bailey (New York: McGraw-Hill, 2000); Don Coursey, "The Demand for Environmental Quality" (St. Louis: John M. Olin School of Business/Washington University, 1992); Seth W. Norton, "Property Rights, the Environment, and Economic Well-Being," in *Who Owns the Environment?* ed. Peter J. Hill and Roger E. Meiners (Lanham, Md.: Rowman and Littlefield, 1998), 37–54; Gene M. Grossman and Alan B. Krueger, "Economic Growth and the Environment," *Quarterly Journal of Economics* 110 (May 1995): 353–377; and John M. Antle and Gregg Heidebrink, "Environment and Development: Theory and International Evidence," *Economic Development and Cultural Change* 43 (April 1995): 603–625.

11. Fernand Braudel, *The Structures of Everyday Life*, vol. 1 of *Civilization and Capitalism: Fifteenth through Eighteenth Century*, trans. Sian Reynolds (New York: Harper and Row, 1985), 41.

12. The rapid population growth is attributable almost entirely to declining death rates (i.e., rising life expectancy), not to rising birth rates. See Nicholas Eberstadt, "World Depopulation: Last One Out Turn Off the Lights," *Milken Institute Review* 2 (first quarter 2000), 37–48.

13. Computed from Fernand Braudel, *The Structures of Everyday Life*, vol. 1 of *Civilization and Capitalism: Fifteenth through Eighteenth Century*, trans. Sian Reynolds (New York: Harper and Row, 1985), 121; and *Statistical Abstract of the United States*, 1996, table 1105.

14. Computed from Braudel, 1:135.

15. Computed from E. Calvin Beisner, *Prospects for Growth: A Biblical View of Population, Resources, and the Future* (Wheaton, Ill.: Crossway Books, 1990), 127.

16. Computed from Braudel, 1:135. See also Richard J. Sullivan, "Trends in the Agricultural Labor Force"; George W. Grantham, "Agricultural Productivity Before the Green Revolution"; Dennis Avery, "The World's Rising Food Productivity"; and Thomas T. Poleman, "Recent Trends in Food Availability and Nutritional Well-Being," in *The State of Humanity*, ed. Julian L. Simon (New York: Blackwell, 1995).

17. See Michael R. Haines, "Disease and Health through the Ages," in *The State of Humanity*, ed. Julian L. Simon (New York: Blackwell, 1995).

18. Both religious and civil liberty were important themes in the political thought of the seventeenth-century Scottish Covenanters, who carried on Knox's tradition. See John Knox, *On Rebellion*, ed. Roger A. Mason (Cambridge: Cambridge University Press, 1994); George Buchanan, *De Jure Regni Apud Scotos* (1579); Samuel Rutherford, *Lex, Rex* (1644); and Sir James Stewart of Goodtrees, *Jus Populi Vindicatum, or, The Right of the People to Defend Their Lives, Liberty, and Covenanted Religion, Vindicated* (1669).

19. See Julian L. Simon and Rebecca Boggs, "Trends in the Quantities of Education: USA and Elsewhere," in *The State of Humanity*, ed. Julian L. Simon (New York: Blackwell, 1995).

20. See Samuel H. Preston, "Human Mortality throughout History and Prehistory"; and Kenneth Hill, "The Decline of Childhood Mortality," in *The State of Humanity*, ed. Julian L. Simon (New York: Blackwell, 1995).

21. See William J. Hausman, "Long-Term Trends in Energy Prices"; Morris A. Adelman, "Trends in the Price and Supply of Oil"; Bernard L. Cohen, "The Costs of Nuclear Power"; John G. Myers, Stephen Moore, and Julian L. Simon, "Trends in Availability of Non-Fuel Minerals"; H. E. Goeller, "Trends in Nonrenewable Resources"; and Roger A. Sedjo and Marion Clawson, "Global Forests Revisited," in *The State of Humanity*, ed. Julian L. Simon (New York: Blackwell, 1995).

22. See William J. Baumol and Wallace E. Oates, "Long-Run Trends in Environmental Quality"; Derek M. Elsom, "Atmospheric Pollution Trends in the United Kingdom"; and Hugh W. Ellsaesser, "Trends in Air Pollution in the United States," in *The State of Humanity*, ed. Julian L. Simon (New York: Blackwell, 1995).

23. Mikhail Bernstam, "Comparative Trends in Resource Use and Pollution in Market and Socialist Economies," in *The State of Humanity*, ed. Julian L. Simon (New York: Blackwell, 1995), 520.

24. See Ronald Bailey, "Earth Day: Then and Now," *Reason* 31 (May 2000): 23.

25. Indur M. Goklany, "Richer is Cleaner: Long-Term Trends in Global Air Quality," in *The True State of the Planet*, ed. Ronald Bailey (New York: Free Press, 1995), 342–343.

26. Calculated from statistics in *Earth Report 2000: Revisiting the True State of the Planet*, ed. Ronald Bailey (New York: McGraw-Hill, 2000), 291–310.

27. Gregg Easterbrook, *A Moment on the Earth: The Coming Age of Environmental Optimism* (New York: Viking, 1995), 582–585.

28. See Stephen Moore, "The Coming Age of Abundance," in *The True State of the Planet*, ed. Ronald Bailey (New York: Free Press, 1995); Lynn Scarlett, "Doing More with Less: Dematerialization—Unsung Environmental Triumph?" in *Earth Report 2000: Revisiting the True State of the Planet*, ed. Ronald Bailey

(New York: McGraw-Hill, 2000); William J. Hausman, "Long-Term Trends in Energy Prices"; Morris A. Adelman, "Trends in the Price and Supply of Oil"; Bernard L. Cohen, "The Costs of Nuclear Power"; John G. Myers, Stephen Moore, and Julian L. Simon, "Trends in Availability of Non-Fuel Minerals"; and H. E. Goeller, "Trends in Nonrenewable Resources," in *The State of Humanity*, ed. Julian L. Simon (New York: Blackwell, 1995).

29. Riane Eisler, *The Chalice and the Blade: Our History, Our Future* (Cambridge, Mass.: Harper and Row, 1987), 174–175.

30. E. Calvin Beisner, "Anomalies, the Good News, and the Debate over Population and Development: A Review of Susan Power Bratton's Six Billion and More," *Stewardship Journal* 3 (summer 1993): 44–53.

31. Nicholas Eberstadt, "Population, Food, and Income: Global Trends in the Twentieth Century," in *The True State of the Planet*, ed. Ronald Bailey (New York: Free Press, 1995), 14–15.

32. Nicholas Eberstadt, "World Population Prospects for the Twenty-First Century: The Specter of 'Depopulation'?" in *Earth Report 2000: Revisiting the True State of the Planet*, ed. Ronald Bailey (New York: McGraw-Hill, 2000), 64. See also Nicholas Eberstadt, "World Depopulation: Last One Out Turn Off the Lights," *Milken Institute Review* 2 (first quarter 2000): 37–48.

33. On population in general, see E. Calvin Beisner, *Prospects for Growth: A Biblical View of Population, Resources, and the Future* (Wheaton, Ill.: Crossway Books, 1990), and "*Imago Dei* and the Population Debate," in *Where Garden Meets Wilderness: Evangelical Entry into the Environmental Debate* (Grand Rapids, Mich.: Eerdmans Publishing/Acton Institute, 1997); Julian L. Simon, *The Economics of Population Growth* (Princeton: Princeton University Press, 1977), *Population Matters: People, Resources, Environment, and Immigration* (New Brunswick, N.J.: Transaction, 1990), and *The Ultimate Resource 2*, rev. ed. (Princeton: Princeton University Press, 1996); Max Singer, *Passage to a Human World: The Dynamics of Creating Global Wealth* (Indianapolis: Hudson Institute, 1987); Michael Cromartie, ed., *The Nine Lives of Population Control* (Washington, D.C., and Grand Rapids, Mich.: Ethics and Public Policy Center/Eerdmans Publishing, 1995); and Michael B. Barkey, Paul Cleveland, and Gregory M. A. Gronbacher, "Population, the Environment, and Human Capital" (forthcoming).

34. Robert C. Balling, *The Heated Debate: Greenhouse Predictions Versus Climate Reality* (San Francisco: Pacific Research Institute, 1992), 65–69.

35. Roy W. Spencer, "How Do We Know the Temperature of the Earth? Global Warming and Global Temperatures," in *Earth Report 2000: Revisiting the True State of the Planet*, ed. Ronald Bailey (New York: McGraw-Hill, 2000), 25.

36. National Research Council, *Reconciling Observations of Global Temperature Change* (Panel on Reconciling Temperature Observations: National Academy Press, 2000).

37. Corrected in 1999 for anomalies related to orbital drift and other problems discovered in 1998.

38. B. D. Santer et al., "Interpreting Differential Temperature Trends at the Surface and in the Lower Troposphere," *Science* 287 (February 18, 2000): 1228. See also Dian J. Gaffen et al., "Multidecadal Changes in the Vertical Temperature Structure of the Tropical Troposphere," *Science* 287 (February 18, 2000): 1242–1245; and David E. Parker, "Temperatures High and Low," *Science* 287 (February 18, 2000): 1216–1217.

39. "Global Warming Smokescreen," *World Climate Report* 5 (March 13, 2000); www.greeningearthsociety.org/climate/previous_issues/vol5/v5n13/feature. htm.

40. Robert C. Balling, *The Heated Debate: Greenhouse Predictions Versus Climate Reality* (San Francisco: Pacific Research Institute, 1992), 92, 102–103.

41. S. Fred Singer, presentation to the 1997 fall meeting of the American Geophysical Union; www.sepp.org/scirsrch/slr-agu.html. Singer's citations are from: [1] A. Trupin and J. Wahr, "Spectroscopic Analysis of Global Tide-Gauge Sea-Level Data," *Geophysical Journal International* 100 (March 1990): 441–453. [2] D. Bromwich, "Ice Sheets and Sea Level," *Nature* 373 (1995): 18. [3] S. L. Thompson and D. Pollard, "A Global Climate Model (Genesis) with a Land-Surface Transfer Scheme," *Journal of Climate* 8 (April 1995): 732–761. [4] H. C. Ye and J. R. Mather, "Polar Snow Cover Changes and Global Warming," *International Journal of Climatology* 17 (February 1997): 155–162. [5] D. A. Meese et al., "The Accumulation Record from the GISP2 Core as an Indicator of Climate Change throughout the Holocene," *Science* 266 (December 9, 1994): 1680–1682.

42. Sherwood B. Idso, *Carbon Dioxide: Friend or Foe?* (Tempe, Ariz.: ibr Press/Institute for Biospheric Research, 1982), 73–80, esp. 73–74, and *Carbon Dioxide and Global Change: Earth in Transition* (Tempe, Ariz.: ibr Press/Institute for Biospheric Research, 1989), 67–107, esp. 68.

43. Sherwood B. Idso, *Carbon Dioxide and Global Change: Earth in Transition* (Tempe, Ariz.: ibr Press/Institute for Biospheric Research, 1989), 67–107.

44. Sherwood B. Idso, *Carbon Dioxide and Global Change: Earth in Transition* (Tempe, Ariz.: ibr Press/Institute for Biospheric Research, 1989), 108. See also *The Greening of Planet Earth*, video and transcript (Arlington, Va.: Western Fuels Association, 1992), 14; and Dennis Avery, "The World's Rising Food Productivity," in *The State of Humanity*, ed. Julian L. Simon (New York: Blackwell, 1995), 381.

45. Among the more important studies on the benefits of enhanced atmospheric CO_2 to plants and, therefore, to agricultural productivity, see Sherwood B. Idso, *Carbon Dioxide and Global Change: Earth in Transition* (Tempe, Ariz.: ibr Press/Institute for Biospheric Research, 1989), and *Carbon Dioxide: Friend or Foe?* (Tempe, Ariz.: ibr Press/Institute for Biospheric Research, 1982). On global

warming in general, see Robert C. Balling, *The Heated Debate: Greenhouse Predictions Versus Climate Reality* (San Francisco: Pacific Research Institute, 1992); Patrick J. Michaels, *Sound and Fury: The Science and Politics of Global Warming* (Washington, D.C.: Cato Institute, 1992); Patrick J. Michaels and Robert C. Balling, *The Satanic Gases: Clearing the Air about Global Warming* (Washington, D.C.: Cato Institute, 2000); Thomas Gale Moore, *Climate of Fear: Why We Shouldn't Worry About Global Warming* (Washington, D.C.: Cato Institute, 1998); Frederick Seitz, Robert Jastrow, and William A. Nierenberg, *Scientific Perspectives on the Greenhouse Problem* (Washington, D.C.: George C. Marshall Institute, 1989); and S. Fred Singer, *Hot Talk, Cold Science: Global Warming's Unfinished Debate* (Oakland, Calif.: Independent Institute, 1997).

46. "Study Finds No Support for Global Warming Fears," *Los Angeles Times*, March 16, 2000, metro section.

47. Jonathan A. Patz et al., "The Potential Health Impacts of Climate Variability and Change for the United States: Executive Summary of the Report of the Health Sector of the U.S. National Assessment," *Environmental Health Perspectives* 108 (April 2000).

48. Thomas Gale Moore, *Climate of Fear: Why We Shouldn't Worry about Global Warming* (Washington, D.C.: Cato Institute, 1998), 88.

49. Frank Cross, "Paradoxical Perils of the Precautionary Principle," *Washington and Lee Law Review* 53 (1996): 919, and "When Environmental Regulations Kill: The Role of Health/Health Analysis," *Ecology Law Quarterly* 22 (1995): 729–784.

50. Arthur B. Robinson, Sallie L. Baliunas, Willie Soon, and Zachary W. Robinson, "Environmental Effects of Increased Atmospheric Carbon Dioxide"; http://zwr.oism.org/pproject/s33p36.html.

51. See Jonathan H. Adler, ed., *The Costs of Kyoto: Climate Change Policy and Its Implications*, and a video by the same title (Washington, D.C.: Competitive Enterprise Institute, 1997).

52. Patrick J. Michaels, "The Consequences of Kyoto," *Cato Policy Analysis* 307 (Washington, D.C.: Cato Institute, May 7, 1998), 8, 5. "Even the former chairman of the IPPC, Bert Bolin, says that the present plan would, if fully implemented, cut warming 25 years hence 'by less that 0.1 degree C, which would not be detectable,' " Thomas Gale Moore, *Climate of Fear: Why We Shouldn't Worry about Global Warming* (Washington, D.C.: Cato Institute, 1998), 143.

53. John H. Cushman, Jr., "Religious Groups Mount a Campaign to Support Pact on Global Warming," *New York Times*, August 15, 1998, section A.

54. Richard S. Lindzen, "Global Warming: The Origin and Nature of the Alleged Scientific Consensus," *Regulation* 15 (spring 1992): 11.

55. Ibid., 8.

56. The Statement by Atmospheric Scientists on Global Warming and a list of signatories can be accessed at www.sepp.org/statment.html.

57. The Heidelberg Appeal and a partial list of signatories can be accessed at www.heartland.org/perspectives/appeal.htm.

58. The Leipzig Declaration on Climate Change and a partial list of signatories can be accessed at www.sepp.org/leipzig.html.

59. The Global Warming Petition developed by the Oregon Institute of Science and Medicine and a list of signatories can be accessed at www.oism.org/pprojects/s33p37.htm.

60. Robert M. May, "Conceptual Aspects of the Quantification of the Extent of Biological Diversity," in *Biodiversity: Measurement and Estimation*, ed. D. L. Hawksworth (London: Royal Society/Chapman and Hall, 1995), 13–20; Paul R. Ehrlich and Anne Ehrlich, *Extinction: The Causes and Consequences of the Disappearance of Species* (New York: Random House, 1981); John Tuxill and Chris Bright, "Losing Strands in the Web of Life," in *The State of the World 1998*, ed. Lester R. Brown, Christopher Flavin, and Hilary French (New York: W. W. Norton, 1998); Jessica Hellman et al., *Ecofables/Ecoscience* (Stanford, Calif.: Stanford University/Center for Conservation Biology, 1998); and Mark H. Williamson, *Island Populations* (Oxford: Oxford University Press, 1981).

61. Julian L. Simon and Aaron Wildavsky, "On Species Loss, the Absence of Data, and Risks to Humanity," in *The Resourceful Earth*, ed. Julian L. Simon and Herman Kahn (Oxford and New York: Blackwell, 1984), 171–183.

62. Timothy C. Whitmore and Jeffrey A. Sayer, ed., *Tropical Deforestation and Species Extinction* (London and New York: Chapman and Hall, 1992).

63. Ibid., 93–94.

64. Charles C. Mann and Mark L. Plummer, *Noah's Choice: The Future of Endangered Species* (New York: Knopf, 1995), chapter 3.

65. Julian L. Simon and Aaron Wildavsky, "Species Loss Revisited," in *The State of Humanity*, ed. Julian L. Simon (New York: Blackwell, 1995), 346–361; Rowan B. Martin, "Biological Diversity: Divergent Views on Its Status and Diverging Approaches to Its Conservation," in *Earth Report 2000: Revisiting the True State of the Planet*, ed. Ronald Bailey (New York: McGraw-Hill, 2000), 203–236; E. Calvin Beisner, "A Christian Perspective on Biodiversity: Anthropocentric, Biocentric, and Theocentric Approaches to Bio-Stewardship," in *Where Garden Meets Wilderness: Evangelical Entry into the Environmental Debate* (Grand Rapids, Mich.: Eerdmans Publishing/Acton Institute, 1997), 129–146.

66. Ronald Bailey, "Earth Day: Then and Now," *Reason* 31 (May 2000): 25.

67. Gregg Easterbrook, *A Moment on the Earth: The Coming Age of Environmental Optimism* (New York: Viking, 1995), 558–559 (adjusted to reflect today's figures). See also "Issue Brief: Endangered Species Act" at www.cei.org/

EBBReader.asp?ID=728, and "Species Removed from the Endangered Species List (Delisted) through February 1997" at www.nwi.org/EndangeredSpecies/Delistings.html.

68. The complete Endangered Species Act may be accessed at www.nesarc.org/act.htm.

69. Rowan B. Martin, "Biological Diversity: Divergent Views on Its Status and Diverging Approaches to Its Conservation," in *Earth Report 2000: Revisiting the True State of the Planet*, ed. Ronald Bailey (New York: McGraw-Hill, 2000), 205.

70. This section is drawn largely from the work of Peter J. Hill, with his permission, especially from his "Biblical Principles Applied to a Natural Resources/Environment Policy," in *Biblical Principles and Public Policy: The Practice*, ed. Richard Chewning (Colorado Springs: NavPress, 1991), 169–182; "Can Markets or Government Do More for the Environment?" in *Creation at Risk? Religion, Science, and Environmentalism*, ed. Michael Cromartie (Washington, D.C.: Ethics and Public Policy Center/Eerdmans Publishing, 1995); and "Takings and the Judeo-Christian Land Ethic: A Response," *Religion and Liberty* 9 (March/April 1999): 5–7. Other studies indicating the importance of private property and free markets to environmental protection include Bernard J. Frieden, *The Environmental Protection Hustle* (Cambridge, Mass.: MIT Press, 1979); Terry L. Anderson and Donald R. Leal, *Free Market Environmentalism* (San Francisco: Pacific Research Institute, 1991); Terry L. Anderson, ed., *Multiple Conflicts over Multiple Uses* (Bozeman, Mont.: Political Economy Research Center, 1994); Elizabeth Brubaker, *Property Rights in the Defense of Nature* (London and Toronto: Earthscan/Environment Probe, 1995); John A. Baden and Douglas S. Noonan, ed., *Managing the Commons*, 2nd ed. (Bloomington: Indiana University Press, 1998); Timothy D. Terrell, "Property Rights and Externality: The Ethics of the Austrian School," *Journal of Markets and Morality* 2 (fall 1999): 197–207; and Michael B. Barkey, "Translating Environmental Ethics into Public Policy," *Journal of Markets and Morality* (forthcoming).

71. For a discussion of the American West and how different property rights systems affected stewardship practices, especially as these practices pertained to species preservation, see Terry L. Anderson and Donald R. Leal, *Free Market Environmentalism* (San Francisco: Pacific Research Institute, 1991), chapter 3.

72. The argument that an adequate information and incentive structure is necessary for good choices to result does not imply that only external incentives and information are all that matter in acting responsibly. As discussed earlier, the value structure of the individual is also crucial, and it is difficult to imagine a well-functioning property rights system without an adequate moral base.

73. In some cases, moral constraints are so strong that they override the badly structured incentives of common property. This usually occurs when the group is small and there is a deep level of commitment to one another and to a shared ideology. For instance, families, local churches, and certain clubs have elements of common property and yet are quite stable over long periods of time. Thus, not all common property arrangements are doomed to failure.

74. See, for instance, Michael Maloney and Bruce Yandle, "Bubbles and Efficiency: Cleaner Air at Lower Cost," *Regulation* 4 (May/June 1980): 49–52; and Michael Levin, "Statutes and Stopping Points; Building a Better Bubble at EPA," *Regulation* 9 (March/April 1985): 33–42.

75. Richard L. Stroup, "The Endangered Species Act: A Perverse Way to Protect Biodiversity," *PERC Viewpoints*, April 1992, page 1. See also Richard L. Stroup, "Endangered Species Act: Making Innocent Species the Enemy," *PERC Policy Series*, April 1995.

76. Randy T. Simmons, "Fixing the Endangered Species Act," in *Breaking the Environmental Policy Gridlock*, ed. Terry L. Anderson (Stanford, Calif.: Hoover Institution Press, 1997), 82.

77. Excerpt from a speech by Jerry Taylor delivered on February 4, 1997, to the Environmental Grantmakers Association: "Environmentalism in a Market Society: Creative Ideas." Mr. Taylor adds, "That's certainly true if a pollutant is truly harmful or a significant nuisance, since individuals—not government authorities—would have the final say over how much pollution they were willing to tolerate on their property or person. That approval would also have the benefit of allowing an array of voluntary contractual relationships between polluter and polluted, internalize the cost of pollution (the holy grail of environmental economics), and minimize the transaction costs and inefficiencies caused by politicized rulemaking."

THE ACTON INSTITUTE

Founded in 1990, the Acton Institute for the Study of Religion and Liberty is named in honor of John Emerich Edward Dalberg Acton, first Baron Acton of Aldenham (1834–1902), the historian of freedom. The mission of the Institute is to promote a free society characterized by individual liberty and sustained by religious principles. To this end, the Institute seeks to stimulate dialogue among religious, business, and scholarly communities. The Institute seeks to familiarize those communities, particularly students and seminarians, with the ethical foundations of political liberty and free-market economics. It also serves as a clearinghouse of ideas for entrepreneurs interested in the ethical dimensions of their vital economic and commercial activities.

The Institute seeks to advance the cause of liberty by working with religious leaders and business professionals to promote the moral and economic dimensions of freedom. Lord Acton understood that liberty is "the delicate fruit of a mature civilization" and that in every age the progress of religious, economic, and political liberty is challenged, even threatened, by its adversaries. Likewise, in our own age, liberty is under constant siege. It is our hope that by demonstrating the compatibility of religion, liberty, and free economic activity, religious leaders and entrepreneurs can forge an alliance that will serve to foster and secure a free and virtuous society.

ACTON INSTITUTE
161 Ottawa Avenue NW, Suite 301
Grand Rapids, MI 49503 USA

GOD SAW ALL
THAT HE HAD
MADE, AND
BEHOLD, IT WAS
VERY GOOD.
(GEN. 1:31)

ACTON INSTITUTE

161 Ottawa Ave NW, Suite 301
Grand Rapids, MI 49503 USA

WWW.ACTON.ORG

PHONE: 616.454.3080

FAX: 616.454.9454

CHECK BELOW FOR YOUR COMPLIMENTARY SUBSCRIPTION

Please send my 1 year complimentary subscription of:

☐ **Acton Notes**   ☐ **Religion & Liberty**   ☐ **Both**

I Prefer: ☐ Mail ☐ Email

PLEASE SEND MY SUBSCRIPTIONS TO:

Name: _____

Position: _____

Organization: _____

Address: _____

City, St, Zip: _____

Email: _____